SOME TRUER METHOD

Sir Isaac Newton in 1720, at age 77.

Engraving by W. T. Fry, after the Kneller portrait. By permission of the Master and Fellows of Trinity College, Cambridge.

SOME TRUER METHOD

REFLECTIONS ON THE HERITAGE OF NEWTON

Edited by
Frank Durham and
Robert D. Purrington

COLUMBIA UNIVERSITY PRESS • NEW YORK

Columbia University Press
New York   Oxford
Copyright © 1990 Columbia University Press

Library of Congress Cataloging-in-Publication Data

Some truer method : reflections on the heritage of Newton /
edited by Frank Durham and Robert D. Purrington.
p.   cm.
Includes bibliographical references.
ISBN 0-231-06896-4
1. Physics—History.
2. Science—Philosophy.
3. Newton, Isaac, Sir, 1642–1727. Principia.
I. Durham, Frank, 1935–
II. Purrington, Robert D.
QC7.5.S65   1990
530'.09—dc20
90-1832
CIP

Casebound editions of Columbia University Press books are
Smyth-sewn and printed on permanent and durable acid-
free paper

Printed in the United States of America

c   10   9   8   7   6   5   4   3   2   1

CONTENTS

CONTENTS

ACKNOWLEDGMENTS

The authors are especially endebted to the Louisiana Endowment for the Humanities and to the National Endowment for the Humanities for support of the Newton Symposium "Newton's Legacy," which provided the stimulus for this work. The editors also would like to acknowledge helpful conversations with Henry Folse and with Jack Wisdom.

1

Newton's Legacy

Robert D. Purrington and Frank Durham

The epitaph on Wren's tomb in the crypt of St. Paul's Cathedral in London, *"Si monumentum requiris, circumspice,"* might equally well be applied to his countryman and contemporary Isaac Newton, for in a profound way, we need but look around us to see the heritage of Newton. The complex web of science and technology which characterizes modern life is thoroughly Newtonian, however roughly Newton may have been treated by the twentieth century. To most of us the world is rational, mechanical, predictable. When entirely new phenomena come to light, they are expected to fit easily under the broad reach of the Newtonian paradigm. We speak, of course, of the much-heralded "Newtonian world-view," surely the greatest of Newton's legacies to our time. But does the term "Newtonian" retain a clear meaning three hundred years after the *Principia* and does the Newtonian *method* continue to have relevance? If indeed our world-view is Newtonian, how did it come to be so, and how have the intervening centuries shaped Newtonianism? What has been the toll of the relativity and quantum theories? These questions and others like them form the *raison d'être* for this volume and for the *Principia* symposium that, for the most part, was its direct antecedent.

In April 1687 Newton, agonized and reluctant, gave up the manuscript of Book III of the *Principia* to Edmund Halley. It had been three years (a period described magnificently by Richard S. Westfall in *Never at Rest*) in which he did little else but struggle with that great work, which ever since has substantially determined how science has been conducted. It was a work which would establish Newton as an almost God-like figure, the very epitome of rationality and profundity; a work whose fame would spread so thoroughly that it would become a model even for the social sciences.

The *Principia* has, nonetheless, been described as the least read of all the important documents of the intellectual history of man. Even during the early eighteenth century influential figures like Locke and Voltaire spread its message without having read or understood it in detail; Locke took Huygens' word on the validity of Newton's arguments. This fate is not unique to the *Principia* among the great scientific works of the past; it was just so with Ptolemy's *Almagest*. Although hundreds of scholars, over fourteen centuries, read and understood it, millions more have found themselves under its influence. Such also was the fate of Copernicus' *De Revolutionibus*, which eventually pushed aside the *Almagest*. Each of these works had sections, usually in the beginning— but in the case of the *Principia*, including parts of Book III as well—which could be understood by the layman. Most, of course, relied on the authority of those who could appreciate the technical aspects of the works.

For a number of reasons, Newton's *Opticks* was, and indeed is, far more widely read than the *Principia*. One factor in this is the somewhat misleading accessibility of the *Opticks*. But the result, for good or ill, was essentially two distinct Newtonian traditions, as Professor Cohen pointed out some years ago. One of the strong influences of the *Opticks* was to lend the weight of Newton's authority to arguments in favor of the existence of atoms and to the interpretation of heat as due to the motion of these particles (query 8).

The influence of the *Principia* is, in any case, clear enough. It greatly influenced Locke, whose work was seen as a philosophical companion to the *Principia*. Coupled with Locke's mechanistic philosophy the two strains of Newtonianism led, respectively, to the analytical triumphs of Euler, Laplace, and Lagrange, among others, and to the nineteenth-century atomistic descriptions of heat and electromagnetism. The success of the techniques of the *Principia* gave credibility to the mechanistic philosophy. The great edifice of British (especially Scottish) physics in the nineteenth century was Newtonian to the core, in spite of the problem of action-at-a-distance, although for some, such as Franklin and Faraday, the model was the *Opticks*, not the *Principia*. Late eighteenth- and early nineteenth-century ideas about electromagnetism and heat came to be expressed in terms of imponderable fluids, whose direct antecedents were Newton's aetherial speculations in the *Opticks*—especially at the hands of Laplace and the school of Arcueil. The extension of Newton's method from gravitation to electricity was mainly qualitative and non-mathematical; Hendry and others have given excellent analyses of the confrontation between the Laplacian mechanistic approach of imponderable fluids and molecular action-at-a-distance, carried forward by Biot and Poisson, and the dynamistic theories of Lagrange, Fresnel, and Fourier.

In other sciences, Newton was a hero to both Lyell and to Darwin (see the chapter by Beatty in this work). Indeed, the three of them are buried within a stone's throw of each other in the nave of Westminster Abbey.

As one reviews the influence of Newton, there emerges a sense that the close of each century since the publication of the *Principia* has seen profound changes in the course of physics. While the skeptic may argue that this is nothing more than a *fin de siècle* phenomenon, or observe that in a process of continual change and evolution—in this case the course of science—such coincidences are likely, even inevitable. Still, it is true that as the eighteenth century drew to a close the influence of the Enlightenment—which drew much of its vitality directly from the example of Newton's success—was at its peak, especially in France. The French and American Revolutions with their emphasis on natural rights grew out of this movement and can thus be seen as a part of Newton's legacy; some of the ideas which pushed forward the American Revolution were brought from France by the ardent Newtonian Benjamin Franklin. Locke had offered an atomistic socioeconomic analysis which emphasized individual rights and a supply and demand economy; in the last decade of the century some of these ideas were carried into the streets.

A century later, to pursue the argument a bit further, classical physics seemed to some a completed science; each of the fields of what we now call classical physics had reached a stage of some maturity, and had taken on much of its present shape. But it was again to be a time of dramatic change, as cultural forces in Europe, which sought release after mid-century, transformed the face of that continent, and brought about an intellectual revolution as well. Indeed, there may never have been, before or since, such a sweeping intellectual revolution as the one that occurred between 1870 and 1914, and which, infused with the terrible lessons of World War I, broke forth again in the decade after the war. Poetry, painting, sculpture, music, and the novel were all transformed; and so, of course was physics.

The problems in classical physics that spawned the quantum theory were lurking near the surface in the physics of the 1880s and 1890s. These included blackbody radiation, atomic and molecular spectra, specific heats, and radioactivity. Planck and Einstein introduced the quantum in the first five years of the new century and Bohr launched the quantum theory of the atom in the following decade. The decade and a half preceding the War are the years of the evolution of the theory of relativity. By 1925 quantum theory had fully emerged. For the first time in two hundred years the Newtonian view of the world was shaken to the core.

And today, just over three hundred years after the *Principia*, the revolutions of relativity and the quantum have been absorbed, at least in the scientific culture. Newton's heritage is more obvious than ever in the accelerated technological remaking of nature, and yet it is at the same time threatened as never before by developments in the physics of the submicroscopic world and in cosmology. This paradox, that the twentieth century is at once the most Newtonian of the centuries that have followed the *Principia*, and the least Newtonian, results from the effectiveness of the Newtonian method (to which, of course, the names of Descartes and others could be added) in fostering the characteristic twentieth-century rational, materialistic world-view. This has occurred in the face of the fundamental alterations of the Newtonian view of space, time, and matter which marked the great revolutions of the first quarter of the century. Add to this the enormously expanded and deepened factual content of the scientific disciplines and one must marvel at the fact that the world continues to be describable, to a remarkable degree, in Newtonian terms.

A question that inevitably arises in any consideration of Newton's impact on this century is, therefore, the extent to which contemporary physics is "post-Newtonian," or even "contra-Newtonian." One response is that while these crucial episodes in twentieth-century physics represent a dramatic departure from classical physics in substance and content, they are, in another, perhaps deeper sense, philosophically Newtonian. We have, it may be argued, a situation in which the twentieth-century world-view is thoroughly Newtonian even though the physics of Newton has been found not to provide a correct or even useful description of a wide range of phenomena. The fundamental view that the universe is a rationally intelligible system, explicable in terms of, and reducible to, a small set of basic laws, survives substantially unchallenged.

It may be, on the other hand, that as the *Principia's* third century comes to a close, we are indeed on the threshold of a truly anti-Newtonian transformation, one which would grow out of attempts to understand the nature of quantum reality. It is fascinating that some of these questions arise from the old problem of action-at-a-distance, which was so important in Newton's own time, but which he effectively sidestepped, and which seemed to have been settled by Faraday and Maxwell in the last century, and later by Feynman et al. in this century.

New awareness of the implications of widespread nonlinearity in nature seems also to be clearly lessening the relevance of the strictly linear Newtonian paradigm, even in areas where Newtonianism would have not long ago have seemed secure—in practical, classical applications to ecology, medicine, meteorology, and so on.

This volume hopes to shed some light on these and other issues that arise in considering Newton from the perspective of the last decade of the twentieth century, three hundred years after the publication of the first edition of the *Principia*. As the title *Some Truer Method* implies, the Newtonian method and its evolution is subjected to scrutiny, especially by Cohen, Westfall, Hankins, and Beatty. But because Newton's heritage is wider than just *method*, and indeed questions arise concerning the continued influence of Newton, the remainder of this volume addresses somewhat more general issues perhaps more philosophical than historical.

We are fortunate to be able to draw on the wisdom of such able and thoughtful scholars as I. Bernard Cohen and Richard S. Westfall, who have written on Newton's century and on Newton's own contributions; Thomas Hankins, who provides a view of Newtonianism from within the growing scientific community of the late eighteenth century; John Beatty, who writes on the reception of Darwin's work into a thoroughly Newtonian world; Joseph Agassi, whose interest is in the problems which arise when one theory bids to replace another, exemplified by the Einsteinian revolution; and Henry Stapp, who discusses current attempts to move beyond the now standard Newtonian materialism. The implications of strong nonlinearity for the interpretation of Newtonianism in the late twentieth century is the subject of the chapter by the editors of this volume.

Clearly the issues addressed here do not encompass all the problems surrounding Newton and the Newtonian revolution that continue to be fiercely debated among historians and philosophers of science, and, with respect to the late twentieth-century problems, among scientists as well. But the tercentenary of the *Principia* has provided the occasion to challenge some long-held ideas, to raise new issues, to renew discussion of insights that need wider attention, and to reappraise some questions which will not go away.

I. Bernard Cohen is the author of the standard *variorum* edition of the *Principia*, with the late Alexander Koyré, and of numerous other works dealing with the Newtonian revolution. Cohen's theme here is one that he has discussed in *Franklin and Newton* and more recently in *The Newtonian Revolution*. It is the dual heritage of Newton for the scientific method, a point we encountered above in noting the parallel influence of the *Principia* and the *Optics*. While questioning the reality of a "Newtonian method" *per se*, Cohen distinguishes a Newtonian "style," which is robust, if difficult to pin down. In order to illustrate the power of the Newtonian style and its relevance for scientists since Newton's time, Cohen goes directly to the *Principia*, where examples of Newton's way of working out problems too broad for any but his approach are to be found.

In an elegant paper that also draws heavily on the *Principia* and on Galileo's early work as well, Richard S. Westfall, author of the definitive scientific biography of Newton, examines the way in which a quantitative, increasingly mathematical view of the world that emerged during the seventeenth century. Concentrating primarily on terrestrial physics, he shows the quantitative gradually replacing the ancient qualitative, and primarily verbal, description that was common before Galileo. In a historical treatment that is nicely complementary to the more philosophical insights of Cohen, Westfall demonstrates the role of Newton in this transformation and the way Newton made the new quantitative tradition his own. Anyone who has forgotten not only Newton's statements about experimental philosophy, but also his devotion to careful and ingenious experimentation, will gain much from Westfall's analysis. He concentrates on the development of that basic device of physics, the pendulum, and suggests that without the pendulum there might have been no *Principia*.

Thomas Hankins is noted for his work on eighteenth-century science, especially French science; in his paper "Newton's 'Mathematical Way' a Century After the *Principia*," he examines the Newtonian tradition at that crucial juncture. By 1787 the work of Euler, d'Alembert, and others had made mechanics a branch of mathematical analysis. For Lagrange force was excluded from the description, and Newton's geometrical arguments had disappeared. Experimentation, becoming ever more quantitative and precise, had been carried into areas far from those countenanced in either the *Principia* or the *Opticks*. Hankins explores the extent to which these and other changes in the first century of Newtonian science constituted a continuation of Newton's method, or a departure from it. To a considerable degree, this question turns on the meaning of the terms "analysis" and "synthesis" as they were employed in the late eighteenth century. Hankins' perspective as a historian in examining these terms, which have played an important role in the philosophy of science, makes it possible to demonstrate usages that do not always conform to the traditional view of the role of synthesis/analysis in the development of modern science.

The ancient contest between mechanists and the proponents of teleology is the subject of John Beatty's paper. Beatty, a leading philosopher and historian of biology, takes as his context Darwinism, which is usually seen as having banished the last vestiges of teleology from science. As Beatty shows, the case is by no means so simple, either in the late nineteenth century or at present. He describes the attempts by both mechanists and teleologists—each including spokesmen who invoked the tradition of Newton to justify their particular position—to claim Darwin as their own. In the process Beatty explores the issues of chance, design, and purpose as they were seen by biologists and commentators. To many, of course, Darwin was Newton's counterpart in biology, and indeed he seems to have seen himself in that light. He was championed by the reductionists of his century, just as he is now embraced by the anti-reductionist biologists of the late twentieth century.

Joseph Agassi, in a paper that is philosophical as well as historical, as befits his far-ranging interests in the philosophy, history, and sociology of science, discusses the status of Newtonian physics in the critical years just before the revolution of special relativity. Was Newtonian physics superseded by the work of Einstein? Clearly not totally: it is in some sense a close approximation to the "better" theory; it is "substantially correct," and may be useful, or even preferable, in some instances. Agassi is concerned with the truth value of theories that have been shown to be "incorrect"; with how one measures the extent to which one theory approximates another or approximately represents the empirical results, which themselves are approximate only in a different sense. How is the choice between necessarily approximate theories to be made when the validity of rival theories cannot be judged against a "true" theory? The late nineteenth century ushered in such an interval, with the aether drift controversy and finally the theories of Einstein. It is Agassi's contention that Einstein's position concerning scientific truth was as radical as his physics, and that the transition to the modern era required Einstein's approach in the one as much as the other.

While the final two chapters focus more directly on the present status of the Newtonian program, they examine very different issues. The first, by the editors of this volume, looks at Newton's heritage in light of recent discoveries concerning strong nonlinearity and chaos in dynamical systems. In particular, the loss of predictability for even relatively simple dynamical systems is used as the point of departure. This discussion is framed entirely within classical physics and deals with three related issues: (1) what, in retrospect, are the implications of chaos for our understanding of Newton's choice of problems in the *Principia* and for the way in which his methods gained ascendancy?; (2) do the discoveries in nonlinear dynamics relegate Newtonianism as it has been understood to an adequate description for only a few rare and special systems?; and (3) what of chaos and determinism?

Henry Stapp's paper, on the other hand, treats overtly post-Newtonian questions of a kind about which Stapp has written widely. He raises, among other matters, the sense in which quantum theory can be treated outside the Newtonian tradition, in its narrow sense. Specifically Stapp discusses the possibility of and the utility of a quantum ontology that would supplement the mathematical structure of quantum theory and provide a basis for interpreting the quantum theory in a much richer way than is possible in the positivist, almost instrumentalist intepretation of Neils Bohr. The most startling conclusion of recent experiments by Aspect et al., which grow out of the famous Einstein, Rosen, Podolsky experiment (to which the name of David Bohm is now usually added), is that the universe is fundamentally and utterly nonlocal. That there are open problems in quantum theory should not be news to anyone— except perhaps to students of quantum mechanics in our graduate schools. The imposition of the Bohr "ontology" and the dogmatic insistence on its validity has arguably proved to have been useful to physics in that it diverted attention from problems of interpretation to applications. Yet, sixty years later, the problems of interpretation, of "meaning," remain unaddressed, and now there are experiments which sharpen the discussions surrounding these issues.

There are reasons for being skeptical of the survival of quantum theory in its present form. These include the fact that quantum theory is an outgrowth of a preoccupation with integrable systems; the fact that such systems are actually rather special and rare (see our chapter in this volume), suggests that a reformulation may be necessary. Furthermore, the linearity of quantum theory is at once its central idea and its greatest weakness. Ultimately it will be necessary to reconcile it in a fundamental way with the general theory of relativity, or its generalization, and the general theory is explicitly nonlinear.

Each of the essays stands alone, of course. Viewed as opportunities seized for timely consideration of the Newtonian heritage, they illustrate important historical and philosophical questions.

It is interesting to note, however, the way in which similar questions arise in differing contexts and across the three centuries since Newton laid down the foundations of modern science. The reader will notice these common issues and contexts as he digests the several chapters. It is not surprising that two of the essays in this volume deal extensively not only with the Newtonian way of doing science, but also with *Newton's* way. As Cohen is at pains to establish, Newton's own approach was more "style" than "method," in the sense that for crucial problems in celestial dynamics, the formal deductive method often was made to stand aside in favor of an all-out attack of a kind that still seems remarkably modern. Westfall's detailed examination of the rise of precise measurement in dynamics during the seventeenth century can be seen as delineating in a fresh way one reason that modern scientists are closer to Newton's style than to his formally announced method.

Similarly, in his discussion of the complex role of analysis and synthesis in eighteenth-century science, which had by then grown substantially beyond the areas that drew Newton's attention, Hankins describes the confusion some scientists of that era felt when they tried to carry Newton's rules from the *Opticks* into practice. Hankins finds what amounts to confirmation of both Cohen's and Westfall's theses in the work of such luminaries as Laplace, Lavoisier, and

Volta. On the one hand, for all that analysis was raised to a status it never had in the *Principia*, the execution of difficult experiments with electricity and heat ultimately relied, Hankins finds, on a combination of experiment and analysis along the lines of the Newtonian "style" delineated by Cohen. On the other hand, the evidence that precision in apparatus and in experimental technique served as driving forces to the development of theory—demonstrated so lucidly in Westfall's paper for the seventeenth century—is found even more strongly in the physics of a century later by Hankins. Indeed, because precise and delicate instruments were by that time being used to characterize quantitatively the intrinsic properties of matter—such as temperature and capacitance, as opposed to extrinsic motions—the concepts had to wait for the instruments, as Hankins discusses. It was at this point, he notes, when analysis was austere and experimentation arcane, that the seed of romanticism sown by Newton began to sprout.

In tracing the contradictory claims made on behalf of Darwinism in the nineteenth century, that Darwinism supports purpose in nature or that it eliminates it, Beatty deals with the problem of the ontological status of a scientific construct. He also finds the view that evolution informs or gives meaning to the molecular or physical-chemical processes that underlie it—rather than the other way round, as in the reductionist position—to be the subject of heated debate in recent years. Of course the terms have changed; teleology is, if not out of the picture, at least well below the surface. Quite surprisingly the fleshing out of a "quantum ontology" as described by Stapp deals with something of the same dichotomy, and with some of the same language. A recovery of "meaning" is seen as occurring when a higher-order description—the non-collapsing wave-function of the universe—is given priority over a microscopic description, the local wave-functions of atoms and molecules. That the universe thereby recovers "creation" and is rescued from being "stillborn" is a resort to biological imagery that partakes more of simile than congruence. Still, the complementary interaction between the mechanistic and the organic ways of characterizing the problems is evident.

One other example of an implicit connection between discussions of apparently disparate aspects of the Newtonian tradition will suffice here. Agassi's concern is with the truth value assigned to theories by observers who are not simply instrumentalists. He calls attention in particular to what in his view amounts to a social need of assurance that the current best descriptions of nature are true, however many generations of theory they themselves may have overturned. He naturally chooses, as one of the two dramatic threats to Newtonian science, the rise of Einstein's ideas, which are often taken as coming from outside the mainstream, but which Agassi sees as profoundly conservative, albeit daring in their lack of claim to finality. We examine in our paper the possibility that a threat to the truth value of Newtonian physics can arise, not from outside the Newtonian tradition, but from Newton's physics itself. The problem in this case is with overlooked or inaccessible mathematical features of the theory. The possibility is raised that a comprehensive theory—indeed Newtonian physics as Newton practiced it—might not survive the improved understanding of the mathematical structure of the theory. This at least serves to reinforce Agassi's sociological point, whether or not one accepts the necessity of a metaphysical commitment to realistic science.

Taken separately or together, the issues that are raised, and we hope clarified, by these essays speak to the vitality of the continuing debate about Newton and the legacy he has left us. Perhaps it cannot be expected that the closing of the present century, the third century of Newton, will bring a new series of revolutionary insights about nature and our place in it, comparable to those of the beginning of this century, for example. We can, however, express the hope, following Newton's remarks in his preface to the *Principia*, that such discussions as this one "will afford some light either on this or some truer method."

2

Newton's Method and Newton's Style

I. Bernard Cohen

1. INTRODUCTION:
THE NEWTONIAN ACHIEVEMENT

Who was this man Isaac Newton? What was the nature of the revolution that he produced in science? In what sense was the revolution so profound that a century later it figured prominently in political thought? We get some measure of Isaac Newton's greatness when we learn that he was the author of not one, but at least two—and maybe even three or four—great revolutions. One was in mathematics and the others in physical sciences. In fact, Newton made so many different kinds of fundamental contributions to science that even if we were to ignore most of them, we would still have to rank him as one of the ten or twelve greatest scientists who ever lived.

In mathematics he invented (along with Leibniz) the calculus—the differential and the integral calculus—the language of the exact sciences and social sciences. He pioneered the use of infinite series and he introduced the method of calculation by successive iterations which is used today in computers. This is his first revolution. In optics he analyzed sunlight and discovered the nature of color. This work led to an understanding of why the sky is blue and to an analysis of the formation of rainbows. Newton used his discovery to invent the reflecting telescope, the principle used in all large telescopes which are built today. Newton's *Opticks* (1704 and later editions) concluded with a general research program in experimental physical science, which formed an agenda for scientists in the eighteenth century.[1] This constitutes a second revolution.

Newton codified the science of mechanics. Those who have studied physics know the three laws of motion, which are fundamental to that subject. It was Newton who invented the modern concept of mass, essential to our study of matter. He also recognized that there are two different measures of mass, which we today call gravitational and inertial, and that an experiment is needed to show their equivalence. This equivalence is a basic feature of Einsteinian relativity theory, but the problem of the equivalence, including the need for a proof, was one of Newton's great discoveries. His establishment of the modern science of "rational mechanics" (Newton's name for this subject) on mathematical principles is the core of Newton's third and greatest revolution.

In Newton's day, however, and for at least a century or more afterward, the great Newtonian Revolution was symbolized by his discovery of the principle of universal gravity. He found the quantitative law of gravity and used it to elaborate the Newtonian "system of the world," to explain the motions of planets, satellites, and comets—the constituent bodies of the universe in which we live. This constitutes a fourth revolution. Generally speaking, historians mean by "the" Newtonian revolution the codification and elaboration of rational mechanics and the development of the system of the world based on gravitational celestial mechanics.

What I find particularly impressive about Newton's genius is that all of this tremendous achievement in science and mathematics was the fruit of only a very small part of his creative life. By profession he was a university teacher; later he became a Member of Parliament. Then, during most of his mature years, he was a public servant, directing the Mint during the recoinage in England in the last decade of the seventeenth century and the opening decades of the eighteenth.

Even during those early creative years for science, when he was a university professor, Newton's main concerns were not exclusively science as we understand that term today— i.e., physics or astronomy or mathematics. Rather, his concerns embraced the interpretation of Scripture, biblical chronology, prophecy, and chemistry and alchemy. We stand in awe before his mighty scientific achievements, the product of only a relatively small part of his creative energy. Even a modest portion of these achievements would have sufficed to earn him an unquestioned place among the scientific immortals.

2. THE NEWTONIAN METHOD

A feature of Newton's achievement was his attention to method. The attempts to codify method—by such diverse figures as Descartes, Bacon, Huygens, Hooke, Boyle, and Newton—signify that discoveries were to be made by applying a new tool of inquiry (a *novum organum*, as Bacon put it) that would direct the mind unerringly to the uncovering of nature's secrets. The new method was largely experimental, and has been said to have been based on induction; it was also quantitative and not merely observational. Therefore it could lead to mathematical laws and principles. I believe that the seventeenth-century evaluation of the importance of method was directly related to the role of experience (experiment and observation) in the new science. For it seems to have been a tacit postulate that any reasonably skilled man or woman should be able to reproduce an experiment or observation, provided that the report of that experiment or observation was given honestly and in sufficient detail. A consequence of this postulate was that anyone who understood the true methods of scientific inquiry and had acquired the necessary skill to make experiments and observations could have made the discovery in the first instance—provided, of course, that he had been gifted with the wit and insight to ask the right questions.

This experimental or experiential feature of the new science shows itself also in the habit that arose of beginning an inquiry by repeating or reproducing an experiment or observation that had come to one's attention through a rumor or an oral or written report. When Galileo heard of a Dutch optical invention that enabled an observer to see distant objects as clearly as if they were close at hand he at once set himself to reconstructing such an instrument. Newton relates how he had bought a prism "to try therewith the celebrated *Phaenomena* of *Colours.*" From that day to this, woe betide any investigator whose experiments and observations could not be reproduced, or which were reported falsely; this attitude was based upon a fundamental conviction that nature's occurrences are constant and reproducible, thus subject to universal laws. This twin requirement of preformability and reproducibility imposed a code of honesty and integrity upon the scientific community that is itself yet another distinguishing feature of the new science.

The empirical aspect of the new science was just as significant with respect to the results achieved as with respect to the aims and methods. The law of falling bodies, put forth by Galileo, described how real bodies actually fall on this earth—due consideration being given to the difference between the ideal case of a vacuum and the realities of an air-filled world, with winds, air resistance, and the effects of spin. Some of the laws of uniform and accelerated motion announced by Galileo can be found in the writings of certain late medieval philosopher-scientists, but the latter (with a single known exception of no real importance) never even asked whether these laws might possibly correspond to any real or observable motions in the external world. In the new science, laws which do not apply to the world of observation and experiment could have no real significance, save as mathematical exercises. This point of view is clearly enunciated by Galileo in the introduction of the subject of "naturally accelerated motion," in his *Two New Sciences*. Galileo states the aim of his research to have been 'to seek out and clarify the definition that best agrees with that [accelerated motion] which nature employs." From his point of view, there is nothing "wrong with inventing at pleasure some kind of motion and theorizing about its consequent properties, in the way that some men have derived spiral and conchoidal lines from certain motions, though nature makes no use of these [paths]." But this is different from studying motion in nature, for in exploring phenomena of the real external world, a definition is to be sought that accords with nature as revealed by experience:

> But since nature does employ a certain kind of acceleration for descending heavy things, we decided to look into their properties so that we might be sure that the definition of accelerated motion which we are about to adduce agrees with the essence of naturally accelerated motion. And at length, after continual agitation of mind, we are confident that this has been found, chiefly for the very powerful reason that the essentials successively demonstrated by us correspond to, and are seen to be in agreement with, that which physical experiments [*naturalia experimenta*] show forth to the senses.[2]

Galileo likens his procedure to having "been led by the hand to the investigation of naturally accelerated motion by consideration of the custom and procedure of nature herself."

Like Galileo, Newton the physicist saw the primary importance of concepts and rules or laws that relate to (or arise directly from) experience. But Newton the mathematician could not help but be interested in other possibilities. Recognizing that certain relations are of physical significance (as that "the periodic times are as the $^3/_2$ power of the radii," or Kepler's third law), his mind leaped at once to the more universal condition (as that "the periodic time is as any power

R^n of the radius R "). Though Newton was willing to explore the mathematical consequences of attractions of spheres according to any rational function of the distance, he concentrated on the powers of index 1 and -2 since they are the ones that occur in nature: the power of index 1 of the distance from the center applies to a particle within a solid sphere and the power of index -2 to a particle outside either a hollow or a solid sphere. It was his aim in the *Principia* to show that the abstract or "mathematical principles" of the first two books could be applied to the phenomenologically revealed world, an assignment which he undertook in the third book. To do so, after Galileo, Kepler, Descartes, and Huygens, was not in itself revolutionary, although the scope of the *Principia* and the degree of confirmed application could well be so designated and thus be integral to the Newtonian revolution in science.

An excessive insistence on an out-and-out empirical foundation of seventeenth-century science has often led scholars to exaggerations. The scientists of that age not did demand that each and every statement be put to the test of experiment or observation, or even have such a capability, a condition that would effectively have blocked the production of scientific knowledge as we know it. But there was an insistence that the goal of science was to understand the real external world, and that this required the possibility of predicting testable results and retrodicting the data of actual experience, the accumulated results of experiment and controlled observation. This continual growth of factual knowledge garnered from the researches and observations made all over the world, paralleled by an equal and continual advance of understanding, was another major aspect of the new science, and has been a distinguishing characteristic of the whole scientific enterprise ever since.

In optics, the science of light and colors, Newton's contributions were outstanding. But his published work on "The Reflections, Refractions, Inflexions [i.e., diffraction] & Colours of Light," as the *Opticks* was subtitled, was not revolutionary in the sense that the *Principia* was. Perhaps this was a result of the fact that the papers and book on optics published by Newton in his lifetime do not boldly display the mathematical properties of forces acting (as he thought) in the production of dispersion and other optical phenomena, although a hint of a mathematical model in the Newtonian style is given in passing in the *Opticks* and a model is developed more fully in section 14 of book one of the *Principia*. Newton's first published paper was on optics, specifically on his prismatic experiments relating to dispersion and the composition of sunlight and the nature of color.

These results were expanded in his *Opticks*, which also contains his experiments and conclusions on other aspects of optics, including a large variety of what are known today as diffraction and interference phenomena. By quantitative experiment and measurement he explored the cause of the rainbow, the formation of "Newton's rings" in sunlight and in monochromatic light, the colors and other phenomena produced by thin and thick "plates," and a host of other optical effects. He explained how bodies exhibit colors in relation to the type of illumination and their selective powers of absorption and transmission or reflection of different colors. The *Opticks*, even apart from the queries, is a brilliant display of the experimenter's art, where (as E. N. de C. Andrade put it so well) we may see Newton's "pleasure in shaping."[3] Some of his measurements were so precise that a century later they yielded to Thomas Young the correct values, to within less than 1 percent, of the wavelengths of light of different colors. Often cited as a model of how to perform quantitative experiments and how to analyze a difficult problem by experiment, Newton's studies of light and color and his *Opticks* nevertheless did not create a revolution and were not ever considered as revolutionary in the age of Newton or afterward. In this sense, the *Opticks* was not epochal.

From the point of view of the Newtonian revolution in science, however, there is one very significant aspect of the *Opticks*: the fact that Newton developed in it the most comprehensive public statement he ever made of his philosophy of science or of his conception of the experimental scientific method. This methodological declaration has, in fact, been a source of some confusion ever since, because it has been read as if it applies to all of Newton's work, including the *Principia.* The final paragraph of query 28 of the *Opticks* begins by discussing the rejection of any "dense Fluid" supposed to fill all space, and then castigates "Later Philosophers" (i.e., Cartesians and Leibnizians) for "feigning Hypotheses for explaining all things mechanically, and referring other Causes to Metaphysicks." Newton asserts, however, that "the main Business of natural Philosophy is to argue from Phaenomena without feigning Hypotheses, and to deduce Causes from Effects, till we come to the very first Cause, which certainly is not mechanical." Not only is the main assignment "to unfold the Mechanism of the World," but also to "resolve" such questions as: "What is there in places almost empty of Matter . . . ?" "Whence is it that Nature doth nothing in vain; and whence arises all that Order and Beauty which we see in the World?" What "hinders the fix'd Stars from falling upon one another?" "Was the Eye contrived without Skill in Opticks, and the Ear without Knowledge of Sounds?" or "How do the Motions of the Body follow from the Will, and whence is the Instinct in Animals?" In query 31, Newton states his general principles of analysis and synthesis, or resolution and composition, and the method of induction:

> As in Mathematicks, so in Natural Philosophy, the Investigation of difficult Things by the Method of Analysis, ought ever to precede the Method of Composition. This Analysis consists in making Experiments and Observations, and in drawing general Conclusions from them by Induction, and admitting of no Objections against the Conclusions, but such as are taken from Experiments. or other certain Truths. For Hypotheses are not to be regarded in experimental Philosophy. And although the arguing from Experiments and Observations by Induction be no Demonstration of general Conclusions; yet it is the best way of arguing which the Nature of Things admits of, and may

be looked upon as so much the stronger, by how much the Induction is more general.

Analysis thus enables us to

> proceed from Compound to Ingredients, and from Motions to the Forces producing them; and in general, from Effects to their Causes, and from particular Causes to more general ones, till the Argument end in the most general.

This method of analysis is then compared to synthesis or composition:

> And the Synthesis consists in assuming the Causes discover'd, and establish'd as Principles, and by them explaining the Phaenomena proceeding from them, and proving the Explanations.

The lengthy paragraph embodying the foregoing three extracts is one of the most often quoted statements made by Newton, rivaled only by the concluding General Scholium of the *Principia*, with its noted expression: *Hypotheses non fingo*.

Newton would have us believe that he had himself followed this "scenario"; first to reveal by "analysis" some simple results that were generalized by induction, thus proceeding from effects to causes and from particular causes to general causes; then, on the basis of these causes considered as principles, to explain by "synthesis" the phenomena of observation and experiment that may be derived or deduced from them, "proving the Explanations." Of the latter, Newton says that he has given an "Instant . . . in the End of the first Book" where the "Discoveries being proved [by experiment] may be assumed in the Method of Composition for explaining the the Phaenomena arising from them." An example, occurring at the end of book one, part 2, is propositions 8–11, with which part 2 concludes. Proposition 8 reads: "By the discovered Properties of Light to explain the Colours made by Prisms." Propositions 9–10 also begin: "By the discovered Properties of Light to explain . . .," followed (proposition 9) by "the Rain-bow" and (proposition 10) by "the permanent Colours of Natural Bodies." Then, the concluding proposition 11 reads: "By mixing coloured Lights to compound a beam of Light of the same Colour and Nature with a beam of the Sun's direct Light."

The formal appearance of the *Opticks* might have suggested that it was a book of synthesis, rather than analysis, since it begins (book one, part 1) with a set of eight "definitions" followed by eight "axioms." But the elucidation of the propositions that follow does not make explicit reference to these axioms, and many of the individual propositions are established by a method plainly labeled "The PROOF by Experiments." Newton himself states clearly at the end of the final query 31 that in books one and two he has "proceeded by . . . Analysis" and that in book three (apart from the queries) he has "only begun the Analysis." The structure of the *Opticks* is superficially similar to that of the *Principia*, for the *Principia* also starts out with a set of definitions (again eight in number), followed by three axioms (three "axiomata sive leges motus"), upon which the propositions of the first two books are to be constructed (as in the model of Euclid's geometry). But then, in book three of the *Principia*, on the system of the world, an ancillary set of so-called "phenomena" mediate the application of the mathematical results of books one and two to the motions and properties of the physical universe. Unlike the *Opticks*, the *Principia* does make use of the axioms and definitions. The confusing aspect of Newton's stated method of analysis and synthesis (or composition) in query 31 of the *Opticks* is that it is introduced by the sentence "As in Mathematicks, so in Natural Philosophy," which was present when this query first appeared (as query 23) in the Latin *Optice* in 1706, "Quemadmodum in Mathematica, ita etiam in Physica." A careful study, however, shows that Newton's usage in experimental natural philosophy is just the reverse of the way "analysis" and "synthesis" (or "resolution" and "composition") have been traditionally used in relation to mathematics, and hence in the *Principia*—an aspect of Newton's philosophy of science that was fully understood by Dugald

Stewart a century and a half ago but that has not been grasped by present-day commentators on Newton's scientific method, who would even see in the *Opticks* the same style that is to be found in the *Principia*.

Newton's "method," as extracted from his *dicta* rather than his *opera*, has been summarized as follows by Colin Turbayne: "The main features of Newton's method, it seems are: The rejection of hypotheses, the stress upon induction, the working sequence (induction precedes deduction), and the inclusion of metaphysical arguments in physics."[4] Thus Turbayne would have "the deductive procedure" be a defining feature of Newton's "mathematical way" and Descartes's *more geometrico* respectively: Descartes' "long chains of reasoning" were deductively linked. Newton's demonstrations were reduced to "the form of propositions in the mathematical way." He would criticize those analysts who have not recognized that the defining property of "the Cartesian 'geometrical method' or the Newtonian 'mathematical way'—paradoxical as it may seem—need be neither geometrical nor mathematical. Its defining property is demonstration, not the nature of the terms used in it."

As we shall see below, however, there is a middle ground between a study of physical or even metaphysical causes and the mathematical elucidation of their action and properties. The recognition of this hierarchy and the exploration of the properties of gravity as a cause of phenomena (without any overt commitment to the cause of gravity) was a great advance over the physics of Galileo.

Thus in the exact sciences of the seventeenth century we may observe a hierarchy of mathematical laws. First, there are those deduced from certain assumptions and definitions, and which lead to experimentally testable results. If, as in Galileo's case, the assumptions and definitions are consonant with nature, then the results should be verifiable by experience. When Galileo sets forth, as a postulate, that the speed acquired in naturally accelerated motion is the same along all planes of the same heights, whatever their inclination, he declares that the "absolute truth of this postulate will be later established for us by our seeing that other conclusions, built on this hypothesis, do indeed correspond with and exactly conform to experiment." This reads like a classic statement of the hypothetico-deductive method; but it is to be observed that it is devoid of any reference to the physical nature of the cause of the acceleration. Such a level of discourse is not essentially different in its results from another seventeenth-century way of finding mathematical laws of nature without going into causes: by the direct analysis of the data of experiment and observation. This was in all probability Kepler's procedure in finding his third (or "harmonic") law of planetary motion. Other examples are Boyle's law of gases and Snell's law of refraction.[5]

The second level of the hierarchy is to go beyond the mathematical description to some sort of causes. Boyle's law, for example, is a mathematical statement of proportionality between two variables, each of which is a physical entity related to an observable or measurable quantity. Thus the volume (V) of the confined gas is measured by the mercury level according to some volumetric scale, and the pressure of the confined gas is determined by the difference between two mercury levels (h) plus the height of the mercury column in a barometer (h_1). Boyle's experiments showed that the product of V and $h + h_1$ is a constant. The sum $h + h_1$ is a height (in inches) of a mercury column equivalent to a total pressure exerted on and by the confined gas; what is measured directly in this case is not the pressure but a quantity (height of mercury) which itself is a measure of (and so can stand for) pressure. But nothing is said concerning the cause of pressure in a confined gas, nor of the reason why this pressure should increase as the gas is confined into a smaller volume, a phenomenon known to Boyle before he undertook the experiments and which he called the "spring" of the air. Now the second level of hierarchy is to explore the cause of this "spring." Boyle suggested two physical models that might serve to explain this phenomenon. One is to think of each particle being itself compressible, in the manner of a coiled spring or a piece of wood, so that the air would be "a heap of little bodies, lying one upon another, as may be resembled to a fleece of wool." Another is to conceive that the particles are in constant agitation, in which case "their elastic power is not made to depend upon their shape or structure, but upon the vehement agitation." Boyle himself, on this occasion, did not choose to decide between these explanations or to propose any other.[6] But the example does show that in the exact or quantitative sciences of the seventeenth century, there was a carefully observed distinction between a purely mathematical statement of a law and a causal mechanism for explaining such a law, that is, between such a law as a mathematical description of phenomena and the mathematical and physical exploration of its cause.

In some cases, the exploration of the cause did not require such a mechanical model, or explanation of cause, as the two mentioned by Boyle. For example, the parabolic path of projectiles is a mathematical statement of a phenomenon, with the qualifications arising from air resistance. But the mathematical conditions of a parabola are themselves suggestive of causes: for—again with the qualifications arising from air resistance—they state that there is uniform motion in the horizontal component and accelerated motion in the downward component. Since gravity acts downward and has no influence in the horizontal component, the very mathematics of the situation may lead an inquirer toward the physical causes of uniform and accelerated motion in the parabolic path of projectiles. Similarly, Newton's exploration of the physical nature and cause of universal gravity was guided by the mathematical properties of this force: that it varies inversely as the square of the distance, in proportional to the masses of the gravitating bodies and not their surfaces, extends to vast distances, is null within a uniform spherical shell, acts on a particle outside of a uniform spherical shell (or a body made up of a series of uniform spherical shells) as if the mass of that shell (or body made up of shells) were concentrated at its geometric center, has a value proportional to the distance from the center within a uniform sphere, and so on.

Such mathematical specifications of causes are different from physical explanations of the origin and mode of action of causes. This leads us to a recognition of the hierarchy of causes which it is important to keep in mind in understanding the specific qualities of the Newtonian revolution in science. For instance, Kepler found that planets move in ellipses with the sun at one focus, and that a line drawn from the sun to a planet sweeps out equal areas in equal times. Both of these laws encompass actual observations within a mathematical framework. The area law enabled Kepler to account for (or to explain) the nonuniformity of the orbital motion of planets, the speed being least at aphelion and greatest at perihelion. This is on the level of a mathematical explanation of the nonuniform motion of planets. Kepler, however, had gone well beyond such a mathematical explanation, since he had assigned a physical cause for this variation by supposing a celestial magnetic force; but he was never successful in linking this particular force mathematically to the elliptical orbits and the area law, or in finding an independent phenomenological or empirical demonstration that the sun does exert this kind of magnetic force on the planets.[7]

Newton proceeded in a different manner. He did not begin with a discussion of what kind of force might act on planets. Rather he asked what are the mathematical properties of a force—whatever might be its causes or its mode of action, or whatever kind of force it might be—that can produce the law of areas. He showed that, for a body with an initial component of inertial motion, a necessary and sufficient condition for the area law is that the said force be centripetal, directed continually toward the point about which the areas are reckoned. Thus a mathematically descriptive law of motion was shown by mathematics to be equivalent to a set of causal conditions of forces and motions. Parenthetically it may be observed that the situation of a necessary and sufficient condition is rather unusual; most frequently it is the case that a force or other "cause" is but a sufficient condition for a given effect, and in fact only one of a number of such possible sufficient conditions. In the *Principia* the conditions of central forces and equal areas in equal times lead to considerations of elliptical orbits, which were shown by Newton to be a consequence of the central force varying inversely as the square of the distance.

Newton's mathematical argument does not, of course, show that in the orbital motion of planets or of planetary satellites these bodies are acted on by physical forces; Newton shows only that within the conceptual framework of forces and the law of inertia, the forces acting on planets and satellites must be directed toward a center and must as well vary inversely as the square of the distance. But in the hierarchy of causal explanation, Newton's result does finally direct us to seek out the possible physical properties and mode of action of such a centrally directed inverse-square force. What is important in the Newtonian mode of analysis is that there is no need to specify at this first stage of analysis what kind of force this is, nor how it acts. Yet Newton's aim was ultimately to go on by a different mode of analysis from the mathematical to the physical properties of causes (or forces) and so he was primarily concerned with "verae causae," causes—as he said—that are "both true and sufficient to explain the phenomena."

There is a great and wide gulf between the supposition of a set of mathematical conditions from which Newton derives Boyle's law and the assertion that this is a physical description of the reality of nature. As will be explained below, it is precisely Newton's ability to separate problems into their mathematical and physical aspects that enabled Newton to achieve such spectacular results in the *Principia*. And it is the possibility of working out the mathematical consequences of assumptions that are related to possible physical conditions, without having to discuss the physical reality of these conditions at the earliest stages, that marks the Newtonian style, as we now will see.

3. THE NEWTONIAN STYLE

As as been shown, the Newtonian revolution in the sciences did not consist merely of his use of deductive reasoning, or of an external form of argument presented as a series of demonstrations from first principles or axioms. Newton's outstanding achievement was to show how to introduce mathematical analysis into the study of nature in a particularly fruitful way, so as to disclose *Mathematical Principles of Natural Philosophy*, as the *Principia* was titled in full: *Philosophiae naturalis principia mathematica*.[8] Newton not only exhibited a powerful means of applying mathematics to nature but also made use of a new mathematics which he himself had been forging and which may be hidden from a superficial observer by the external mask of what appears to be an example of geometry in the traditional Greek style.

In the *Principia* the science of motion is developed in a way that I have characterized as the Newtonian style. It shall be seen that this style consists of an interplay between the simplification and idealization of situations occurring in nature and their analogues in the mathematical domain. In this manner Newton was able to produce a mathematical system and mathematical principles that could then be applied to natural philosophy, that is, to the system of the world and its rules and data as determined by experience. This style permitted Newton to treat problems in the exact sciences as if they were exercises in pure mathematics and to link experiment and observation to mathematics in a notably fruitful manner. The Newtonian style also made it possible to put to one side, and to treat as an independent question, the problem of the cause of universal gravity and the manner of its action and transmission.

An outstanding feature of Newton's scientific thought is the close interplay of mathematics and physical science. It is, no doubt, a mark of his extraordinary genius that he could exercise such skill in imagining and designing experiments, and in performing such experiments and drawing from them their theoretical significance. He also displayed a fertile imagination in speculating about the nature of matter (including its structure, the forces that might hold it together, and the causes of the interactions between varieties of matter). In the present context, my primary concern is with mathematics in relation to the physical sciences of dynamics and celestial mechanics, and not with these other aspects of Newton's scientific endeavors.

The "principles of natural philosophy" that Isaac Newton displayed and elaborated in his *Principia* are "mathematical principles." His exploration of the properties of various motions under given conditions of force is based on mathematics and not on experiment and induction. What is not so well known is that his essays in pure mathematics (analytic geometry and calculus) often tend to be couched in the language and principles of the physics of motion. This interweaving of dynamics and pure mathematics is also a characteristic feature of the science of the *Principia*. We shall see that Newton shows himself to be a mathematical empiricist to the extent that he believed that both basic postulates and the final results of mathematical analysis based on those postulates could be consonant with the real or external world as revealed by experiment and critical or precise observation. But his goal was attained by a kind of thinking that he declared explicitly to be on the plane of mathematical rather than physical discourse and that corresponds to what we would call today the exploration of the consequences of a mathematical construct or a mathematical system abstracted from, yet analogous to, nature.

Newton's achievement in the *Principia* was, in my opinion, due to his extraordinary ability to mathematicize empirical or physical science. Mathematics at once served to discipline his creative imagination and thereby to sharpen or focus its productivity and to endow that creative imagination with singular new powers. For example, it was the extension of Newton's intellectual powers by mathematics and not merely some kind of physical or philosophical insight that enabled him to find the meaning of each of Kepler's laws and to show the relationship between the area law and the law of inertia. The power of mathematics may also be seen in Newton's analysis of the attraction of a homogeneous sphere. Newton proves that if the force varies either as the distance directly or as the square of the distance inversely, then the gravitational action of the sphere will be the same as if the whole mass of the sphere were to be concentrated at the geometric center. The two conditions (as Newton points out in the scholium to proposition 78, book one) are the two principal cases in nature. The inverse-square law applies to the gravitational action on the surface or at a point outside the sphere (the force inside having been proved to be null). The direct-distance law applies to the action on a particle within a solid sphere. It might have been supposed that in any solid body, the centripetal force (as Newton says) of the whole body would "observe the same law of increase or decrease in the recess from the center as the forces of the particles themselves do," but for Newton this is a result that must be attained by mathematics. Mathematics shows this to be the case for the above two conditions, a fact which Newton observes "is very remarkable."

The main thrust of the present discussion is the way in which Newton's mathematical thought was especially suitable for the analysis of physical problems and the construction and alteration of models and imaginative constructs and systems, but it must be kept in mind that some of Newton's basic mathematical concepts were themselves derived from physical situations. Since Newton tended to think in terms of curves that are drawn or traced out by moving points, his primary independent variable was time. In fact, his discussion of time in his purely mathematical treatises so closely resembles the presentation of time in the *Principia* (under the heading "absolute, true, and mathematical time") that it would be difficult to tell them apart, out of context.

There is an obvious pitfall in making too much of the language (images as well as metaphors) of physics in Newton's mathematics. For, when in his October 1666 tract on fluxions (or calculus), he writes "To resolve Problems by Motion," he is actually concerned with pure mathematics, even though the language may suggest physics; but this would have been true of all writers on "locus" problems since Greek times, who would trace out a curve or a line by a moving point or construct a solid by revolving a plane figure about an axis.[9] Newton's success in analyzing the physics of motion depended to a large degree on his ability to reduce complex physical situations to a mathematical simplicity, in effect by studying the mathematical properties of an analogue of the reality that he eventually wished to understand. Thus we shall see him exploring by mathematics the motion of a mass point in a central force field as a first step toward understanding the significance of Kepler's area law as a general rule and not in relation to any specific orbital system. Newton was quite aware of the difference between the mathematical properties of such simplified analogous constructs and the physical properties expressed in mathematical relations or rules or principles of the physical world as revealed by experiments and observations; but later readers and some scholars today have tended to blur Newton's usually clear distinctions. In formalizing and developing his mathematical principles of natural philosophy, Newton made use of his own new mathematics, although this fact is apt to be masked from the reader by the general absence of a formal algorithm for the calculus in the *Principia*. This new mathematics appears in his early papers both in a purely algebraic or symbolic presentation much as in a present-day treatise on analysis (although with different symbols), and in a discussion of motion from a mathematical point of view. The latter is of interest to us here, because what is at issue is not merely a vague kinematical tracing out of the conditions of a locus, but rather the elaboration for the purposes of pure mathematics of the geometry of curves based on principles of motion that are also used in physical kinematics. What is perhaps even more significant than the close conceptual fit of Newton's pure mathematics to solutions of physical problems is that, while there is a mode of thought common to both his mathematics and physics, there is in the *Principia* a continual awareness of the fundamental difference between mathematical principles and natural philosophy expressed through mathematical principles.

By this I mean that for purely mathematical purposes—that is, in a mathematical context and not with the aim of elucidating problems of physics—Newton uses principles of motion that read just as if they were physical principles being applied to physical local motion (locomotion), including the resolution and composition of vector speeds and the concept of inertial or uniform motion. Care must be exercised, according to D. T. Whiteside's warning, lest we assume too hastily that Newton was using physical principles in pure mathematics. Rather he was constructing a mathematical system that was analogous to (but not identical with) a physical system. That is, his mathematical "time" is not the physical time of experience; and it is the same with respect to mathematical "speed," and so on. Nevertheless, he did use the same language in both his writings on the physics of motion and his development of mathematics by a mathematics of motion. I believe, although of course there can be no proof, that there is a close bond between Newton's tendency to think about pure mathematics in terms that are the same as those arising in the physics of motion and his insight and skill in using pure mathematics to solve problems in physical motion. Yet one should not make too much of such a linkage, which would have been operative only on the subconscious level, since Isaac Barrow (for one) had also written pure mathematics in the language of motion, which may been Newton's direct source of inspiration.

The reader who has never studied Newton's mathematical writings can have no idea of the quasi-physical imagery of motion in his presentation of the method of fluxions. For example, in his *Treatise of the Method of Fluxions and Infinite Series*, he observed that "all the difficulties" may be "reduced to these two Problems only, which I shall propose, concerning a space described by local Motion, any how accelerated or retarded":

I. The length of the Space described being continually (that is, at all times) given; to find the velocity of the motion at any time proposed.

II. The velocity of the motion being continually given; to find the length of the Space described at any time proposed.

If Newton thus conceived of fluxions and limits in terms of a mathematical local motion, it may not be surprising that he should have developed a powerful tool for analyzing local motion in a physical sense by means of mathematics that used the method of limits, as in the *Principia*. Many years later, in about 1714, in a draft of his anonymous book review of the *Commercium epistolicum*, in which the Newtonian priority in the discovery of the calculus is asserted, Newton made it clear once again that mathematical concepts similar to those used in the physics of motion were basic to his own version of the calculus.[10] "I consider time," he wrote, "as flowing or increasing by continual flux & other quantities as increasing continually in time, & from the fluxion of time I give the name of fluxions to the velocities with which all other quantities increase." His method was to "expose time by any quantity flowing uniformly" and, in a manner reminiscent of Galileo, he said that his "Method is derived immediately from Nature her self."[11]

This intimate connection between pure mathematics and the physics of motion is, I believe, a characteristic feature of Newton's *Principia*, wherein certain aspects of natural philosophy are reduced to mathematical principles, then developed as mathematical exercises, and finally reapplied to physical problems. The main subject of the *Principia* is terrestrial and celestial dynamics: the physics of motion, or the motion of bodies under the action of various kinds of forces and different conditions of restraint and resistance, and the mathematical method is fluxional and uses vanishing infinitesimals; it is a characteristic feature to apply the limit process to geometric conditions and to proportions (or equations) arising from or representing those conditions. Hence the quasi-physical nature of Newton's mathematics was eminently suited to the solution of the problems to which he addressed himself in his *Principia*. But while this intermingling of a pure mathematics derived from or related to motion and the physical problems of motion may have led Newton to achieve unheard-of results of astonishing fecundity, this very aspect of his work has caused great confusion among his commentators and interpreters ever since. In particular, they have not always known when Newton was speaking on the level of mathematics and when on the level of physics. Or, they have perhaps assumed this to be a distinction without a difference and have not bothered to ascertain whether Newton as a mathematician was—in the *Principia*—everywhere intending to be understood as a physicist. It will be seen below that a major aspect of Newton's method in the *Principia* (and possibly in other aspects of his work in the exact sciences) was his intuitive separation of these two levels of discourse and then, on the proper occasion, using his mathematical results to illuminate the physical problem. The blurring of Newton's distinctions, which has led to continual misunderstandings concerning Newton's method and his intentions, probably derives from a reading of certain scholia and introductory sections of the *Principia* out of the context of the mathematical physics in which they are embedded and which they were intended to illuminate.

One of the clearest statements Newton ever made of his own position was in reply to a criticism made by Leibniz. The details of this criticism would take us far afield, and we need only take notice here that Newton held that what his critic "saith about Philosophy is foreign to the Question & therefore I shall be very short upon it." Newton's grounds for this statement of disagreement with Leibniz concerning the fundaments of "Philosophy" (i.e., natural philosophy) were threefold. First, "Leibniz denies conclusions without telling the fault of the premisses." Second, "His arguments against me are founded upon metaphysical & precarious hypotheses & therefore do not affect me: for I meddle only with experimental Philosophy." Third, "He changes the signification of the words Miracles & Occult qualities that he may use them in railing at universal gravity . . ." In writing the last sentence, he had at first used the words "railing at me," which shows the degree to which in his own mind he had equated himself with the conceptual fruit of his intellect.

As Newton said again and again, there was a fundamental difference in philosophy between himself and Leibniz. To deny universal gravitation could be legitimate in Newton's philosophy only by going back to the arguments given by Newton, and to the premises of those arguments: a combination of empirical finding, mathematical development, and sound logic. It was not enough merely to say that a concept of universal gravitation is not acceptable in philosophy. And so to understand the fundaments of Newtonian exact science (i.e., the exact science of the *Principia*), it is necessary to see what in fact were the stages by which Newton got to universal gravitation. In so doing, we shall see why Newton held that there is a profound difference between "metaphysical & precarious hypotheses" and "experimental Philosophy." Finally, Newton was particularly concerned about "Miracles & Occult qualities." He stoutly denied the relevance of "Miracles" to his natural philosophy in the sense of a suspension of the ordinary laws of nature, and he equally denied that he had reintroduced into science the "Occult qualities" of late Aristotelian-scholastic philosophy. Gravitation itself was not "Occult," but its cause was, in the degree that it was still hidden from us.

Newton's spectacular achievement in producing a unified explanation of the events in the heavens and on our earth, and in showing how such diverse phenomena as the ebb and flow of the tides and the irregularities in the moon's motion might be derived from a single principle of universal gravity, draw attention to his mode of procedure—a special blend of imaginative reasoning plus the use of mathematical techniques applied to empirical data—which I have called the Newtonian style. Its essential feature is to start out (phase one) with a set of assumed physical entities and physical conditions that are simpler than those of nature, and which can be transferred from the world of physical nature to the domain of mathematics. An example would be to reduce the problems of planetary motion to a one-body system, a single body moving in a central force field; then to consider a mass point rather than a physical body, and to suppose it to move in a mathematical space in mathematical time. In this construct Newton has not only simplified and idealized a system found in nature, but he has also imaginatively conceived a system in mathematics that is the parallel or analogue of the natural system. To the degree that the physical conditions of the system become mathematical rules or propositions, their consequences may be deduced by the application of mathematical techniques.

Because the mathematical system (to use an expression of Newton's in another context) duplicates the idealized physical system, the rules or proportions derived mathematically in one may be transferred back to the other and then compared and contrasted with the data of experiment and observation (and with experiential laws, rules, and proportions drawn from those data). This is phase two. For instance, the condition of a mass point moving with an initial component of inertial motion in a central force field is shown (*Principia*, propositions 1 and 2, book one) to be a necessary and sufficient condition for the law of areas, which had been found to be a phenomenologically verifiable relation in the external astronomical world.

The comparison and contrast with the reality of experiential nature (that is, with the laws, rules, and systems based upon observations and experiments) usually require a modification of the original phase one. This leads to further deductions and yet a new comparison and contrast with nature, a new phase two. In this way there is an alternation of phases one and two leading to systems of greater and greater complexity and to an increased *vraisemblance* of nature. That is, Newton successively adds further entities, concepts, or conditions to the imaginatively constructed system, so as to make either its mathematically deduced consequences or the set conditions conform more exactly to the world of experience. In the example under discussion, the first of these additional steps is to introduce Kepler's other laws of planetary motion. The third law, applied to uniform circular motion in combination with the Newtonian (Huygenian) rule for centripetal (centrifugal) force yields the inverse-square force, as does a parabolic or hyperbolic orbit.

In the next stage of complexity or generality, Newton adds to the system a second body or mass point, since (as Newton says at the start of section 11, book one, of the *Principia*) attractions are not made toward a spatial point but rather "toward bodies," in which case the actions of each of the bodies on the other are always equal in magnitude though oppositely directed. Yet additional conditions include the introduction of bodies with finite sizes and defined shapes, and of a system of more than two interacting bodies. (There is also the question of whether bodies move through mediums according to some specified law of resistance.)

For Newton there is a final stage in this process: when the system and its conditions no longer merely represent nature simplified and idealized or an imaginative mathematical construct, but seem to conform to (or at least to duplicate) his realities of the external world. Then it becomes possible, as in book three of the *Principia*, to apply the aggregate of mathematical principles to natural philosophy, to elaborate the Newtonian system of the world. This is the final phase three of the Newtonian style, the crown of all, to display the variety of natural phenomena that can be attributed to the action of universal gravity. It is not until after this stage that Newton himself would have to yield to the demand for investigation into the nature, cause, or mode of operation of such forces as he had used in accounting for the motions of terrestrial bodies, the planets, their moons, our moon, the comets, the tides, and various other phenomena. This additional inquiry, a kind of sequel to phase three, went beyond the requirements of the Newtonian style, however, at least insofar as the *Principia* is concerned. Even in the General Scholium, with which the later editions of the *Principia* conclude, Newton insisted that his gravitational dynamics and his system of the world could be accepted even though he had said nothing about the cause of gravity. But he did then express his personal conviction that gravity "really exists."

It is a feature of the Newtonian style that mathematics and not a series of experiments leads to the most profound knowledge of the universe and its workings. Of course, the data of experiment and observation are used in determining the initial conditions of the inquiry, the features that yield the mathematical principles that are applied to natural philosophy, and Newton was also aware that the success of the eventual natural philosophy (or of the system of the world) must rest ultimately on the accuracy or validity of the empirical data of which it was constructed. Furthermore, the test of the end result was necessarily the degree and extent of the ability to predict and to retrodict observed phenomena or phenomenologically determined "rules" (such as Kepler's laws). Even so, on some significant occasions Newton seems to have given priority to the exactness of mathematical system rather than the coarseness of empirical law. In the case of Kepler's laws, the reason is that they prove in Newton's analysis to be exact only in a very limited situation and to be no more than phenomenologically "true" (that is, "true" only within certain conventionally acceptable limits of observational accuracy) of the real world as revealed by experience. Hence, even the Newtonian system of the world, when said by Newton (in the final edition of the *Principia*) to be based on "phenomena," is in fact based also to some degree on truths of mathematical systems or idealizations of nature that are seen to be approximate to but not identical equivalents of the conditions of the external world.

The simplified physical system (and its mathematical analogue that Newton develops at the beginning of the *Principia*) exhibits all three of Kepler's laws and in fact serves to explain these laws by showing the physical significance of each one separately. In short, this system or construct is not a figment of the free imagination, or a purely arbitrary or hypothetical fiction created by the mind, but is rather closely related to the real Copernico-Keplerian world that is made known to us by phenomena and by laws that are phenomenologically based. In his first flush of victory, just before writing the *Principia*, when he had completed his analysis of this system, Newton himself thought that it was more than an imagined construct. Indeed, in a first version he expressed his belief that he had now explained exactly and fully how nature works in the operation of the solar system. But not for long. It was almost at once plain that the construct he had been studying did not accord with the real world. And so, bit by bit, it was endowed with more and more features that would bring it into closer harmony with the world of reality. In the course of these transformations of his construct, Newton was led by degrees to the concept of a mutual gravitating force, a concept all the more conspicuous by its absence from the first considerations. As a result, it is possible to assign a precise limiting date to the first step toward that great concept: no earlier than December 1684.

The advantage of the Newtonian approach, as I have outlined it, are manifold. First of all, by making the construct simple at the start, Newton escapes the complexities of studying nature herself. He starts out with an idealized version of nature, in which certain descriptive laws of observed position and speeds—Kepler's planetary laws—hold exactly. Then, on the basis of the laws and principles that underlie these descriptive laws, Newton proceeds to new constructs and to more general underlying laws and principles, and eventually gets to the law of universal gravity in a new system in which the original three planetary laws as stated by Kepler are—strictly speaking—false.

Is Newton's ultimate system still only an imagined construct? Or is it now so congruent with reality that its laws and principles are the laws and principles of the universe? Newton does not tell us what he believes on this score, but we may guess at his position. His first construct, for which Kepler's laws are valid, proved to be a one-body system: essentially a single mass-particle moving under the action of a force directed toward a fixed center. Next he extended and modified the results found true of the one-body system so that they would apply to a two-body system, in which each of the two bodies acts mutually on the other with the same inverse-square force that acts on the single body in the first one. Then there are many bodies, each acting on all the others mutually with an inverse-square force; and, finally, the bodies have physical dimensions and determinate shapes, and are no longer mere mass points or particles. The force is shown to duplicate gravity and to be mutually acting, and is then found to be a universal force proportional to the product of the masses. In this way, Newton extended his construct from one to two mass points and then to many, and from particles or mass points to physical bodies. Since there are no more bodies to be added, I believe he would have argued that the system was complete. Further physical complexities or mathematical conditions might be imagined and put in: for instance, gross bodies with negative masses; or bodies that would interact with other bodies, but that might have negative gravitational forces (repulsions) as well as positive ones (attractions), just in electrical and magnetic phenomena. But in terms of accumulated observations over many centuries, any such condition would then have been ruled out as highly improbable, if not downright impossible. Of course, since Newton was unable to give a general solution to the problem of three mutually gravitating bodies, there might have been unforeseen complications in a many-body system. But we may guess that these speculative conditions would not have carried much weight with him. He had found the system of the world.

Furthermore, the final system would certainly have seemed to have transcended the status of being merely an imagined construct in the degree to which its results agreed with many different kinds of observations. The Newtonian theory could explain not only why all bodies fall with the same acceleration at a given place on earth, but also the observed fact that the acceleration varies in a certain definite manner with latitude (as shown by the concomitant variation in the period of a freely swinging simple pendulum) and other factors. The theory of gravity could also explain the tides and many features of the motion of the moon, and could even predict the oblate shape of the earth from the known facts of precession. The variety and exactness of the verifiable predictions and retrodictions of experience gave every reason to believe that the Newtonian system of the world, displayed in book three of his *Principia*, and developed and extended by others, was indeed the true system of the world. And so it was conceived for over two hundred years—until Einstein developed his theory of relativity.

Because the final system achieved by Newton was found to work so well, it no longer had to be treated as an imagined construct. As Newton himself put it in the general scholium, there are three conditions for gravity that are enough, that suffice in natural (or experimental) philosophy. It is enough ("Satis est"), first of all, "that gravity really exists"; second, that gravity "acts according to the laws that we have set forth"; third, that gravity "is sufficient to explain all the motions of the heavenly bodies and of our sea." For Newton there then arose two wholly different sets of questions. The first were technical: to work out, as he saw, the "details" of gravitational celestial mechanics and thus to get better results for such problems as the motion of the moon. This range of activity can be described as completing the *Principia* on the "operative" level. The second set of questions was of another type altogether: to explain gravity and its mode of action or to assign "a cause to gravity." His critics, however, proceeded in just the opposite manner, starting out with vexing problem of how such a force as the proposed Newtonian universal gravity can possibly exist and act according to the Newtonian laws, and not accepting the formal results of the *Principia* so long as they did not find its conceptual basis to be satisfactory. These critics, in other words, were not willing to go along with the procedural mode of the Newtonian style.

4. THE SIGNIFICANCE OF THE NEWTONIAN STYLE AND METHOD

The Newtonian achievement, of course, did not consist wholly of the introduction into science of the use of imagined systems and mathematical constructs as found in the *Principia*. Rather that revolution was a radical restructuring of the principles and concepts of motion along the lines of mass, acceleration, and forces; plus the elaboration of a system of the world operating in terms of the new dynamics, in which universal gravity is the governing force and inertia is a primary or essential property of matter. In terms of both the breath of its scope and profundity of the analysis, the *Principia* was unfolded in 1687 to a wholly unexpectant and unprepared audience who did not know what to make of it or how to use it for some time. Only gradually was it appreciated how deeply Newton had seen into the operations of nature, that he had recognized both the role of mass in inertial physics and the distinction between mass as a resistance to acceleration (what we call inertial mass) and as a determinant of force in a gravitational field (gravitational mass). It is difficult to think of any other scientific book that embodied so complete a change in the state of knowledge concerning the physics of the heavens and the earth.

Newton produced such an astonishing revolution in science by applying mathematics (geometry, algebra or proportions, fluxions, limit procedures, infinite series) to natural phenomena. It was Newton's example that such later scientific figures as Kant and Quetelet had in mind when they averred that a science's progress may be measured by the degree to which it becomes mathematical. Newton's style is of paramount importance because it made possible his mathematicization of nature's processes. And it is in this sense that this style gives a key to the revolution in science associated with the *Principia*.

Newton was far from being the first scientist to build a mathematical system of nature. Ptolemy and Archimedes were distant (but only partial) predecessors; more immediate ones were Copernicus, Galileo, Descartes, and—above all—Kepler; while Huygens and Wallis were older contemporaries. But Archimedes had addressed himself to a very limited range of nature, while Ptolemy's *Mathematical Composition* embraced certain aspects at the expense of others. As a result, Ptolemy's systems for the sun, moon, and planets are essentially a set of geometric computing schemes or geometric models and appear not to have been intended as a representation of reality.

Pierre Duhem has called attention to the problem of devising computing schemes versus attempting to portray reality; he views the conflict between the two as a dominant theme in the history of the science of the heavens from Greek times to the seventeenth century.[12] His thesis, that there was a complete dichotomy down the ages between model-makers and realists, is extreme. Nevertheless, we may see the conflict between these two points of view coming to the fore with Copernicus's *De revolutionibus*.[13] This book had been printed in such a way as to give the impression that the author was presenting his new system as only a model, a hypothesis, or a scheme for computing solar, planetary, and lunar phenomena. Following the dedication to Pope Paul III, there is an introductory essay ('Ad lectorem de hypothesibus hujus operis'), in which Copernicus appeared to have said just that. Kepler found evidence, however, that this essay was not a composition by Copernicus at all; it had been introduced into the book by Andreas Osiander (a Protestant clergyman who saw *De revolutionibus* through the press). Kepler himself, as we have seen, was primarily a realist. He wanted to start right out with the causes of the motions of the planets and to find their true paths; he was not interested in any mere computing schemes. And so he was especially delighted to find that Copernicus also had been a realist, that Copernicus had believed heart and soul in his own system and had not been the author of the prefatory essay on hypothesis.

Galileo, as convinced a Copernican as Kepler, did not either improve the schemes for calculation or seek for causes. Of course he believed in the reality of the Copernican system, and even concocted a reason (based on an explanation of the tides) why there had to be rotation of the earth as well as a motion in revolution around the sun. Galileo did not, however, concern himself notably with technical details of the Copernican system so much as with philosophical and scientific arguments favoring a general heliocentrism and an anti-Aristotelian view of motion. It is therefore in his *Two New Sciences* rather than in his *Two Chief Systems of the World* that we may seek for a true precursorship of the Newtonian style as a step toward applying mathematics to nature. For example, Galileo had to deal with the realities of air friction or air resistance in relation to the motion of pendulums and the free fall of bodies. Since this real situation was too complex and difficult for him to handle, he simplified nature as he found it by supposing a world of empty space in which the air would have no effect. He predicted, for example, that in such an imagined world, a coin and a feather would fall freely at the identical rate, or have equal accelerations. On a smaller scale than Newton's, Galileo was thus considering a simplified case of physics as a stage toward reality. The slight difference in time of fall of a heavy and a light body dropped at the same time from a tower was attributed to air friction, the cause of the divergence of the idealized situation from reality. Pendulum experiments then provided evidence that air friction does indeed resist motion. Only in the case of an idealization or simplification of nature, and not in the real world of ordinary experience, are Galileo's laws of free fall and the parabolic trajectories of projectiles strictly valid.

In the present context, these Galilean examples are of only academic interest—for the record, so to speak—since we have

no reason to believe that Newton had ever read Galileo's
Two New Sciences, and many bits of evidence to indicate
that he had not.[14] Important and fruitful research remains
to be done on the whole question of the use of imagined
systems and mathematical constructs in seventeenth-century physical science, both the systems and constructs that
comprise a set of mathematically expressed conditions of
force, resistance, and motion and those that embody physical systems or mechanisms to explain theories (and are like
the "models" of today's scientists and philosophers of science). Such a study would no doubt point to possible sources
of Newton's procedure, which he would have transformed,
improved, and endowed with extraordinary new powers.

Did Newton ever show any direct awareness that he was
doing something new in his use of the Newtonian style? Not
in so many words. But he was certainly aware that no one
before him had found the many results he had discovered.
He knew, of course, that some earlier scientists had guessed
at the inverse-square law, and he even supposed that this
law might have been known to the ancients. But that did
not mean that he took any the less credit for his invention:
the inverse-square law may have been known to seers of a
bygone day, but to have found it and to have proved that it
caused elliptical orbits was something new in his own age,
and to this degree he would not give Hooke or anyone else
any credit for it. Furthermore, what was significant about
the Newtonian achievement was not merely to guess or even
to know the inverse-square law, but rather to use it to demonstrate elliptical orbits and to develop a system of the world
based upon. This could not have been done by experiment
and observation, by induction, or by philosophical speculation, but only by mathematics. And the key to applying
mathematics to the world of nature was the Newtonian style,
in which by stages there could be added the conditions that
would bring the original imagined system and mathematical construct into congruence with the realities of experience. The mathematics needed for this task was a new
mathematics, the fluxional calculus, embodied in the con-

tinual use of limits and of infinite series, and here Newton insisted that he was the sole inventor—the first inventor, not a rediscoverer of ancient methods.[15]

The degree to which the Newtonian style was revolutionary can be seen in the simple fact that so much of our exact science ever since has proceeded in a somewhat similar manner. I believe that there is a logical and simple penchant for a kind of positivism on the part of all who approach physical subjects as mathematicians, and for whom the exploration of the mathematical consequences of any system or of any set of conditions is equally fascinating; although certain ones are obviously of more importance than others, because they relate to nature as revealed by experiment and observation.

In the general scholium written for the second edition of the *Principia* in 1713, Newton expressed the quasi-positivist view which has inspirited much of the exact sciences from that day to ours, when he said 'it is enough' that gravity exists and that from it we can deduce the motion of the heavenly bodies, terrestrial objects, and the tides. By this expression Newton was, as we have seen, arguing against the kind of criticism in which the whole structure of Newtonian celestial dynamics would be thrown out on metaphysical grounds, because of the abhorrence of "attraction" or doubts as to whether such a force can possibly exist. In 1717, in the second edition of the *Opticks*, Newton once again stated that he did not consider how "these Attractions may be performed."[16] Repeating essentially what he had said in 1706 in the Latin edition, he pointed out, "What I call Attraction may be performed by impulse, or by some other means unknown to me." And he stressed that he was using "that Word [attraction] here to signify only in general any Force by which Bodies tend toward one another, whatsoever be the Cause." He had limited himself to the first stage of inquiry, in which "we must learn from the Phaenomena of Nature what Bodies attract one another, and what are the Laws and Properties of the Attraction, before we enquire the Cause by which the Attraction is performed." But none of these statements can be taken to imply that Newton himself was uninterested in the search for such causes, or that he himself had not instituted such a search.

Newton's insistence that it is enough to be able to predict the celestial and terrestrial motions, the tides of the sea, etc., was in fact less a battle cry of the new science than a confession of failure. For what Newton was saying in essence is that his system should be accepted in spite of his failure to discern the cause or even to understand universal gravity, because its results accord so well with the data of observation and experiment. The eventual acceptance of Newtonian celestial mechanics, in the absence of the knowledge of the fundamental cause, was in one sense a perversion of the philosophy Newton had expressed in the General Scholium, because it inhibited any further search for a cause. But in another sense, modern science *has* been following the Newtonian principles, because Newton *did* believe that the first goal of mathematical physics (or of exact science) is to predict and retrodict the phenomena of nature.

In any case Newton's statements should not be taken to mean that celestial and terrestrial dynamics were in any possible sense "complete" or in a "final form" as they appeared in the *Principia*. Newton had taken a giant step forward, as may be seen by contrasting the level of discourse (the range, generality, complexity, and difficulty) and the actual subject matter of Galileo's *Discorsi* and *Dialogo* and Newton's *Principia*. This is also the case, although to a far lesser degree, with regard to Kepler's *Astronomia nova* or Huygens's *Horologium oscillatorium*. The greatness of Newton's achievement made dramatically clear what he had omitted from his considerations and also what he had done badly, imperfectly, or incompletely. For example, he had opened up the possibility of perturbation theory but had not made a truly significant contribution to it. Newton's stature is hardly diminished by an honest appraisal of his achievements and his failures, by the recognition that giants were still needed after the *Principia* in order to revise, improve, and complete the subjects he had treated and to exercise their collective genius to create all the other subjects that constitute classical exact science.

The scientists of the eighteenth century were well aware of the problems unsolved by Newton as well as those topics which he had pursued with such success. They were as cognizant of Newton's failures as of his impressive breakthroughs. We do not at present have an exact picture of the response to Newtonian science because the impression made by the *Principia* on the science of the eighteenth century has never been fully studied. What is needed is more than an examination of the influence of a general Newtonian philosophy or an inquiry into the stages and degrees of acceptability of the concept of gravitation; on these topics there are many monographic studies of excellent quality. To evaluate the impact of the *Principia* with precision, it is necessary to study in detail and in depth how individual scientists actually used Newtonian principles, methods, laws, concepts, and particular results. In such an inquiry, it does not suffice merely to see whether or not the three Newtonian laws of motion are used as such, although it may be the greatest interest to inquire into whether (or in what form) Newton's second law may appear. It would be particularly fruitful to see how the approach found in the beginning of book one of the *Principia* or in the propositions around proposition 41 may have been used by different writers on dynamics. In some cases the Newtonian principles, concepts, methods, or results will be found to have been adopted without change, but in others it will be found that there had to be significant transformations.

Such an investigation may show that a concept that might otherwise have seemed to us to be not very fruitful was apparently of real service. An instance may be seen in Newton's "vis inertiae." On page 3 of d'Alembert's *Traité de dynamique*, he says: "Following Newton, I call *force of inertia* [*force d'inertie*] the property which bodies have of remaining in the state in which in which they are; now a Body is necessarily in either a state of rest or a state of motion . . ."[17] This leads him to two laws:

1st law
A Body at rest will persist in rest unless an external cause draws it out of that state. For a Body cannot of itself put itself in motion.

2nd law
A Body once put into motion by any cause whatever must ever persist in motion that is uniform and in a straight line, if indeed a new cause, different from the former one which put the body in motion, does not act on it, that is to say, unless an outside force that is different from the cause of the motion acts on this Body, and it will move continually in a straight line and will traverse equal spaces in equal times.

It is thus plain that d'Alembert is not following Newton in having a simple and single law of inertia, but rather—having accepted Newton's "force d'inertie"—he has gone back to the two separate laws of inertia in a form like that used by Descartes in his "leges naturae" in his *Principia philosophiae.*

I believe that the outlook of post-Newtonian scientists, using a system based on the action of a universal force that they could not understand, was not wholly unlike that of present-day physicists with respect to quantum field theory. In a 1977 talk, Murray Gell-Mann said:

All of modern physics is governed by that magnificent and thoroughly confusing discipline called quantum mechanics, invented more than fifty years ago. It has survived all tests and there is no reason to believe that there is any flaw in it. We suppose that it is exactly correct. Nobody understands it, but we all know how to use it and how to apply it to problems; and so we have learned to live with the fact that nobody can understand it.[18]

It was much the same in the age of Newton for gravitational celestial mechanics. After a while, men of science like Euler, Lagrange, and Laplace used such concepts as universal gravity and force because this was the key to the successful mathematical solution of so many problems arising in physics and astronomy. By the time of Laplace's *Méchanique céleste*, Laplace begins with "the motion of a material point" (an obviously Newtonian mathematical construct), but shifts easily to aspects of terrestrial gravity, the motions of projectiles, and other aspects of observable physics.[19]

Of course, there were always physicists who were concerned with the philosophical problems of the nature of force—including such diverse figures as Boscovich, Mach, and Hertz. This too was in a Newtonian tradition, however un-Newtonian their solutions to the Newtonian problem may have been, for it is directly related to what I have called the sequel to phase three of the Newtonian style. Newton's demonstration of the dynamical significance of Kepler's laws and the limits and conditions of their veracity, and his explorations of the motion of the moon, the perturbation of Saturn's motion by Jupiter, his explanations of tidal phenomena in the seas and the motion of comets, the computation of the masses of planets with satellites, and the study of the precession of the equinoxes as a consequence of the pull of the moon on the equatorial bulge of an earth that is an oblate spheroid—all of these and more must have collectively exerted a strong pressure on scientists to pursue the Newtonian system even without necessarily believing in it, without necessarily understanding how its central operative feature, universal gravity, could exist and could act as it must in that system. Happily, a main feature of the Newtonian revolution in science was the Newtonian style, which provided a sharp division between the use of concepts like centripetal force and universal gravity that one could not understand and the search for the causes of such forces or the attempts to understand them.

Despite the impressive achievements of the *Principia*, Newton's failures could not but be evident to any first-rate mathematical physicist. Leibniz, Johann Bernoulli, Euler, Laplace were certainly not (in any simple and direct sense of the word) Newtonians. Even Clairaut, who held that Newton's *Principia* had created a revolution, found it necessary to warn against trusting Newton as our guide, and he made his reputation not so much for his commentary on Newton's treatise as for producing a genuine contribution to the theory of perturbations in which Newton had failed. Laplace, who had the highest praise for Newton's general achievement and who openly expressed his admiration for many parts of the *Principia*, could not help but observe that while Newton was "fort ingénieux" there were many places where he was not "heureux" in either his methods or his results. Newton, furthermore, had ignored certain topics of paramount importance, notably "the mechanics of rigid and flexible and fluid and elastic bodies." Clifford Truesdell finds that book two of the *Principia* offered a challenge to the geometers of the day: "to correct the errors, to replace the guesswork by clear hypotheses, to embed the hypotheses at their just stations in a rational mechanics, to brush away the bluff by mathematical proof, to create new concepts so as to succeed where Newton had failed."[20] He concludes that "rational mechanics, and hence mathematical physics as a whole and the general picture of nature accepted today, grew from this challenge as it was accepted by the Basel school of mathematicians: the three great Bernoullis and Euler," on the basis of whose work a succession of first-rate men during the eighteenth and nineteenth centuries "constructed what is now called classical physics." Furthermore, there are major subjects of mechanics—statics, energy considerations, rigid bodies—for which the *Principia* provided no direct illumination. How then could Newton's *Principia* have been of such primary importance in the eighteenth century?

I believe that the answer to this question is not to be found merely by looking for the ways in which a particular proposition or method was used or rejected or corrected. Nor do I believe that Newton's importance was merely in his use of such concepts as mass, centripetal force, universal gravity. Of course it was impressive to have demonstrated that many phenomena could be accounted for by universal gravity, and to have introduced a theoretical basis for the unification of the phenomena of the heavens and of our earth.

But I believe that his stature and influence must be seen in more general terms than a measure of either his outstanding successes or his dismal failures, in relation to what Truesdell has called the "program" that Newton "laid down for us and illustrated by brilliant examples" in the *Principia*. Of course, the *Principia* contains many beautiful and ingenious solutions of outstanding problems, notably in particle mechanics and dynamical astronomy, and even his only partial successes and his failures set forth new kinds of problems that were not even generally imagined a generation earlier.

But above all Newton set forth a style of science that showed how mathematical principles might be applied to physics and astronomy (that is, to natural philosophy) in a particularly fruitful way. This may have been even more influential in the long run than his system of the world based on universal gravity. But, of course, what gave the Newtonian style, and the book in which it was set forth by example, a more than ordinary importance was not its degree of success so much as the primary nature of the subject to which it was addressed: the system of the world. As Lagrange wistfully remarked, and as Laplace agreed, there was but one law of the cosmos and Newton had discovered it.

The Newtonian style is seen to be more a universal characteristic of the science and its practitioners than of the man Newton. The Newtonian style was not a completely original creation of Newton's, but a transformation in which he brought to a high level a tradition going back to Greek antiquity, one that had been undergoing a series of radical or significant transformations during the seventeenth century. I believe it to be a sign of Newton's genius that he was able to grasp the potentialities of this style and to transform it so effectively in the working out of his mathematical philosophy of nature applied to problems of dynamics and the mechanics of the system of the world.

AUTHOR'S NOTE. The present article largely represents a reworking of material which has appeared earlier, primarily in *The Newtonian Revolution* (New York: Cambridge

I. Bernard Cohen

University Press. 1980. Copyright Cambridge University Press 1980). The material from *The Newtonian Revolution* is reprinted with the permission of Cambridge University Press. The author expresses his gratitude to Frank Durham for his efforts in the adaptation of the present article.

NOTES AND REFERENCES

1. Isaac Newton, *Opticks or a Treatise of the Reflections, Refractions, Inflections & Colours of Light.* Based on the fourth edition: London, 1730. With a foreword by Albert Einstein; an introduction by Sir Edmund Whittaker; a preface by I. Bernard Cohen; and an analytical table of contents prepared by Duane H. D. Roller. New York: Dover Publications, 1952.

2. Galileo Galilei, *Two New Sciences, Including Centers of Gravity and Force of Percussion.* Translated with introduction and notes by Stillman Drake. Madison: The University of Wisconsin Press, 1974. p. 153. Galileo Galilei, *Le Opere di Galileo Galilei.* vols. Editor, Antonio Favaro; *coadiutore letterario,* Isidoro del Lungo; *assistente per la cura del testo,* Umberto Marchesini (Vittorio Lami for vol. 3). Florence: tipografia di G. Barbera, 1890–1909, 8:197.

3. E. N. da C. Andrade, "Newton," in The Royal Society's *Newton Tercentenary Celebrations, 15-19 July 1946.* Cambridge (England): at the University Press, 1947, pp. 3–23; p. 12.

4. Colin Murray Turbayne, *The Myth of Metaphor.* New Haven: Yale University Press, 1962.

5. Ernst Mach, *The Principles of Physical Optics, an Historical and Philosophical Treatment.* Translated by John S. Anderson and A. F. A. Young. London: Methuen & Co., 1926. Reprint, New York: Dover Publications, 1953, pp. 32–36. A. I. Sabra, *Theories of Light from Descartes to Newton.* London: Oldbourne, 1967. Edmund Hoppe, *Geschichte der Optik* Leipzig: Verlagsbuchhandlung J. J. Weber, 1929, pp. 33ff.

6. I. Bernard Cohen, *Franklin and Newton: An Inquiry into Speculative Newtonian Experimental Science and Franklin's Work in Electricity as an Example Thereof.* Philadelphia: The American Philosophical Society, 1956. Reprint, Cambridge: Harvard University Press, 1966. p. 103. Robert Boyle, *The Works of the*

Honourable Robert Boyle. 5 vols. London: printed for A. Millar, 1744. A "new edition" in 6 vols, London: printed for J. and F. Rivington, L. Davis, W. Johnston, 1772, 1:11ff.

7. Alexandre Koyré, *The Astronomical Revolution: Copernicus-Kepler-Borelli.* Translated by R. E. Maddison. Paris: Hermann; London: Metheun; Ithaca (N. Y.): Cornell University Press, 1973. pt. 2, sect. 2, ch. 6. Eric J. Aiton, "Kepler's Second Law of Planetary Motion," *Isis* 60 (1969): 75–90. Curtis Wilson, "Kepler's Derivation of the Elliptical Path," *Isis* 59(1968): 5–25.

8. *Isaac Newton's Philosophiae Naturalis Principia Mathematica.* The third edition (1726) with variant readings assembled by Alexandre Koyré, I. Bernard Cohen, and Anne Whitman. 2 vols. Cambridge (England): at the University Press; Cambridge (Mass.): Harvard University Press, 1972.

9. Derek T. Whiteside, "Patterns of Mathematical Thought in the Later Seventeenth Century," *Archive for History of Exact Sciences* 1 (1961): 179–388. Isaac Newton, *The Mathematical Papers of Isaac Newton.* Vol. 1, 1664–1666 (1967); vol. 2, 1667–1670 (1968); vol. 3, 1670–1673 (1969); vol. 4, 1674–1684 (1971); vol. 5, 1683–1684 (1972); vol. 6, 1684–1691 (1974); vol. 7, 1691–1695 (1976); vol. 8, 1697–1722 (1981). Edited by D. T. Whiteside, with the assistance in publication of M. A. Hoskin and A. Prag. Cambridge (England): at the University Press, 1:369, n. 2.

10. Isaac Newton. "An account of the book entitled *Commercium Epistolicum Collinii & Aliorum, De Analysi Promota,* published by order of the Royal-Society, in relation to the dispute between Mr. Leibniz and Dr. Keill, about the right of invention of the new geometry of fluxions, otherwise call'd the differential method." *Philosophical Transactions* 29(1715): 173–224.

11. Isaac Newton, *The Mathematical Papers of Isaac Newton,* 3:17.

12. Pierre Duhem, *To Save the Phenomena, an Essay on the Idea of Physical Theory from Plato to Galileo.* Translated by Edmund Doland and Chaninah Maschler with an introduction by Stanley L. Jaki. Chicago: The University of Chicago Press, 1969.

13. Nicholas Copernicus, *De Revolutionibus Orbium Coeles-*

tium. Nuremberg: apud. Joh. Petreium, 1543. Facsimile reprint, Paris: M. J. Hermann, 1927.

14. I. Bernard Cohen, "Newton's Attribution of the First Two Laws of Motion to Galileo," in *Atti del Symposium Internazionale di Storia, Metodologia, Logica e Filosofia della Scienza "Galileo nell Storia e nella Filosofia della Scienza."* Collection des Travaux de l'Academie Internationale d'Histoire des Sciences, no. 16. Vinci (Florence): Gruppo Italiano di Storia della Scienza, 1967, pp. xxv–xxliv.

15. John Collins et al., *Commercium epistolicum J. Collins et aliorum de analysi promota, etc., ou, Correspondance de J. Collins et d'autres savants celebres du XVIIe siecle, relative a l'analyse superieure, reimprimee sur l'edition originale de 1712 avec l'indication des variantes de l'edition de 1722, completee par un collection de pieces justificatives et de documents, et publiee par J.-B. Biot et F. Lefort.* Paris: Mallet-Bachelier, 1856. See also Newton, 1715, note 10.

16. Isaac Newton, *Opticks or a Treatise of the Reflections, Refractions, Inflections & Colours of Light*, qu. 31, par. 1, 376.

17. Jean Le Rond d'Alembert, *Traité de Dynamique. . . .* Paris: chez David l'aîné, 1743. Reprinted, Brussels: Culture et Civilisation, 1967.

18. Murray Gell-Mann, "The Search for Unity in Particle Physics," Communication presented at the 1580th stated meeting of the American Academy of Arts and Sciences, April 13, 1977.

19. Pierre Simon, Marquis de Laplace, *Celestial Mechanics*. Translated with a commentary by Nathaniel Bowditch. 4 vols. New York: Chelsea Publishing Company, 1966. Corrected facsimile reprint of the volumes published in Boston in 1829, 1832, 1834, 1839, ch. 2, sect 9–10, bk. one.

20. Clifford Truesdell, "Reactions of Late Baroque Mechanics to Success, Conjecture, Error, and Failure in Newton's *Principia*," in Robert Palter, ed., *The Annus Mirabilis of Sir Isaac Newton 1666–1966*. Cambridge: MIT Press, 1970, pp. 192–232.

3

Making a World of Precision: Newton and the Construction of a Quantitative Physics

Richard S. Westfall

When we look back at the scientific revolution from our perspective of three centuries, seeking to understand that most momentous of "events" in the intellectual life of the West, one characteristic insistently catches our eye—the ever greater role of mathematics and of quantitative modes of thought, what Alexander Koyré aptly called the geometrization of nature. In the seventeenth century, the geometrization of nature confined itself largely to physics. In the eighteenth century, quantitative considerations played a major role in the restructuring of chemistry. In our own era, these considerations have made a major contribution to molecular biology. Therefore to be a scientist today it is necessary to understand and to do mathematics. Indeed, one might well conclude that the geometrization of nature is perhaps our most distinctive legacy from the scientific revolution.

Science was not always so. Before the scientific revolution, through a span of more than two thousand years, Western mankind's understanding of nature was dominated by Aristotle, who had taught an entirely different natural philosophy. This is not the setting for a summary of Aristotelian philosophy, but I will offer two examples from my own recent studies by way of illustration.

Among his papers, Galileo left a manuscript commentary on Aristotle's *De caelo*, which was published in the National Edition of Galileo's works, early in this century, under the heading "Juvenilia."[1] As a result of recent scholarship, we now know that Galileo drew the commentary directly from the lecture notes of professors at the Jesuits' Collegio Romano, and because we can identify the professors, whose periods of instruction at the Collegio are recorded, we also know that far from being a juvenile exercise, the commentary was the work of a mature young man of about twenty-five who was probably already a professor of mathematics at the University of Pisa.[2] The manuscript serves as a nice introduction to natural philosophy as it was pursued before the scientific revolution.

Several characteristics of the discussion provide an instructive contrast to central features of modern science. First, the manuscript looked consistently to authority. In saying this, we must be careful not to repeat without examination Galileo's own later taunts against the Aristotelians. Like Aristotle himself two thousand years earlier, the commentary accepted nature as the ultimate authority in natural philosophy. It sought a philosophical account that would correspond to nature, and we shall be badly misled if we think that it looked for natural philosophy solely in the pages of printed books. Nevertheless, on every question it raised the commentary cited the opinion of Aristotle, which it usually concluded by upholding. Usually, but not always—for the commentary also recognized another authority still higher than Aristotle, to wit, Christianity. When Aristotle conflicted with Christian doctrine, the commentary invariably upheld the Christian position. In this it marked another small chapter in the continuing chronicle of tension that began with the reception of Aristotelian philosophy in the Latin West in the twelfth century. In Galileo's manuscript, the issue was the eternity and perfection of the world. Following his Jesuit mentors, who in turn followed a well established line of argument, Galileo replied that the world is neither eternal nor perfect. Rather it was created in time by an almighty God whose omnipotence is such that He could have created an infinity of other worlds more perfect.

A second feature of the commentary is its verbal character. One looks in vain for any mathematics in it. In one passage, on the intensity of qualities, the manuscript stated that the highest degree of intensity is eight. One must not allow the number to confuse him; even it was purely verbal. It was not the result of any measurement, nor did it look to any operation or instrument by which it could be measured. Rather it was merely a way of talking about intensity. Do not imagine that I intend to disparage the commentary in this respect. Like Aristotelian natural philosophy in general, it was engaged in sorting out the basic concepts necessary for an effective natural philosophy. Thus its discussion of the extension and intensity of qualities effectively distinguished quantity of heat from temperature. Surely one would be ill-advised to make light of such a discussion. I find it impossible to imagine modern science without the background of four centuries of rigorous schooling in Aristotelian natural philosophy. Nevertheless—and this is the point on which I wish to insist—the verbal character of Aristotelian natural philosophy made it an utterly different enterprise from modern science with its quantitative character.

The third aspect of the manuscript follows directly from the second: it thought of nature in qualitative rather than quantitative terms. Thus the four elements it accepted were distinguished by pairs of four basic qualities, wet, dry, hot, and cold. The words did not refer, as they do for us, to particular sensations, but rather to qualities deemed truly to exist in the ontology of nature, which was then qualitative far more than it was quantitative.

Of course the manuscript commentary embodied other features, but the three above are the ones on which I wish to draw in my own discussion. Before I leave the commentary, however, let me cite briefly a typical passage in which one can virtually feel its difference from modern science. The passage considered the proposition that every species has a limit such that no individual member of the species can be bigger. Suppose there were a horse as big as a horse can be, whose owner in fury beat it so that it swelled; would it then cease to be a horse? If the question was typical, so also was the solution of the difficulty by means of a distinction. Limits by way of corruption, the manuscript argued, are external rather than internal; since the swelling would be the result of corruption, the beast would not cease to be a horse. I admit to being fascinated by the discussion, but I cannot avoid the judgment that I would not likely meet a similar one in a modern work on science.

My other example is also associated with Galileo. A quarter of a century after he compiled the commentary above, Galileo was present at a gathering where an Aristotelian philosopher asserted that ice is condensed water. Let us note in passing that it was not a silly assertion but one that followed directly from the principles of his philosophy. Galileo retorted that the proposition was impossible. Ice floats in water, and ice must therefore be rarer than water. The dispute became the occasion of Galileo's second major publication, his *Discourse on Bodies in Water*, and of four Aristotelian pamphlets written in reply to it. The most important of the four pamphlets was composed by Ludovico delle Colombe, who demanded that Galileo explain the power of cold to rarefy. As every philosopher knew, it was the virtue of heat to dilate and open, while the virtue of cold was to compress and restrict. If such is the case, why does ice float in water? Because of its shape. Ice forms in extended sheets on the surface of water. The larger the sheet, the more the dryness of the solid is extended, and hence the greater is the resistance of water (which is in its fluidity contrary to dryness) to being divided. Therefore ice floats.[3]

To this argument based on the virtues of qualities and the resistance of water to division, Galileo replied with the simple quantities of Archimedean hydrostatics. All bodies are heavy. Those with a higher specific gravity will sink when placed in fluids of lesser specific gravity. A solid will float in a fluid of greater specific gravity, however, and if it is forced down, it will be extruded. There is no doubt about which argument sounds more convincing to us. We recognize Galileo's position; we hardly know how to deal with Colombe's. With the exception of a tiny handful of mavericks, the entire world of learning at the beginning of the seventeenth century would have confronted the two in exactly the opposite order.

As everyone knows, Galileo elaborated the methodological stance of the *Discourse on Bodies in Water* into a mathematical conception of nature. A passage in *The Assayer* contains the most familiar statement of it.

> Philosophy is written in this grand book, the universe, which stands continually open to our gaze. But the book cannot be understood unless one first learns to comprehend the language and read the letters in which it is composed. It is written in the language of mathematics, and its characters are triangles, circles, and other geometric figures without which it is humanly impossible to understand a single word of it.[4]

As everyone also knows, this was not just abstract talk on Galileo's part. He embodied it in the kinematics of free fall, the mathematical relations of time, distance, velocity, and acceleration in uniformly accelerated motion.

Moreover, Galileo was not alone. At much the same time, Kepler shared much the same outlook.

> Geometry [said Kepler], being part of the divine mind from time immemorial, from before the origin of things being God Himself (for what is in God that is not God Himself?) has supplied God with the models for the creation of the world and has been transferred to man together with the image of God. Geometry was not received inside through the eyes.[5]

On Kepler's part as well this was not just abstract talk, for he embodied it in the kinematics of planetary motion, the three mathematical laws of orbital motion that we still accept. It would not be difficult to cite others from the years ahead who shared this position. I tend to be wary of terms that call upon the spirit of an age. Nevertheless I find it hard to avoid a similar concept in discussing the early years of modern science. Among a small handful of intellectual leaders (in effect, those whom we remember), near the beginning of the seventeenth century, a new outlook on natural philosophy appeared. For them, suddenly, quantitative arguments began to seem the only solid ones. Among the following generations, others who could respond to the same compulsions learned the new point of view from them. This was the heritage Newton received. In the *Principia* he defined the quantitative dynamics that underlies the kinematics expounded earlier, and in doing so provided the final warrant for the new conception of nature, establishing the character of science that still prevails among us.

Historians of science have documented the appearance of the concept that nature is fundamentally quantitative and mathematical. What I hope to do in this paper is to bring the concept down to a lower plane of generality and to trace some aspects of the development of a quantitative physics that embodied the general philosophic principles. Part of the development of quantitative physics took place within astronomy, the original home of quantitative science. In the generation before Galileo and Kepler, Tycho Brahe illustrated the coming new outlook. There was nothing in Tycho's observations that would have been technically impossible several centuries earlier. In fact, however, it was only at the end of the sixteenth century that anyone found that degree of quantitative precision worth pursuing. Before the end of the seventeenth century, other astronomers equipped with telescopic sights would greatly exceed Tycho's level of precision. There would have been no *Principia* and no modern science of the sort we know without that body of precise astronomical observations.

It is my intention, however, to pursue terrestrial physics (understand quantitative terrestrial physics) in this paper rather than celestial. Already I have introduced one example, Archimedian hydrostatics. When Galileo spoke of specific gravity, he did not mean something like the Aristotelian intensity of qualities that he had employed in his early commentary on *De caelo*. He could measure specific gravity on real instruments. Because of the economic importance of the precious metals, he had precision balances at his command, and he knew how to employ them to measure specific gravity. Thus he could issue a challenge to Colombe: did the resistance of water to division that Colombe employed in his argument exist apart from floating bodies? Could Colombe measure it?[6] For Galileo could measure specific gravity apart from the phenomenon of floating bodies. In addition to quantities measured on a balance, Galileo could also measure lengths. But beyond the two dimensions of weight and length (and associated quantities such as specific gravity) there was precious little pertaining to terrestrial physics that Galileo or any other scientist could measure at the beginning of the seventeenth century. The core of my story is the success of seventeenth-century science in creating a quantitative physics without the abundant supply of instruments we associate with science today—or, perhaps another way to saying the same thing, the success of seventeenth-century science in creating instruments sufficient for a quantitative physics.

The very heart of the new natural philosophy was mechanics, the science of motion. Mechanics required the measurement of a third dimension, time. The creation of the new world of precision was intimately connected to the success of science in learning to measure time. The effort began with Galileo. His concept of uniformly accelerated motion entailed the conclusion that distances traversed from rest must vary as the square of the time. Could he test that consequence? Free fall proceeded far too swiftly, but Galileo understood that he could dilute the rate of acceleration by using an inclined plane. He cut a groove in the edge of a

board, lined it carefully with parchment in order to mini-
mize resistance, and rolled a bronze ball down the groove.
To measure time he used a water clock, a deep container
with a small hole in the bottom. He caught the water in a
container and measured elapsed time by weighing the water.[7]
Recently a historian of science repeated the experiment as
Galileo described it and found that he could consistently
achieve an accuracy within one or two percent of the anticipated result.[8]

For the purposes of this experiment, Galileo needed only
to compare different periods of time to each other, for example, the times required to traverse one, four, and nine
feet. If his procedure was adequate to compare periods, it
was incapable of measuring time in the already established
units of hours, minutes, and seconds with similar precision.
But Galileo needed this measurement as well. He had introduced a fundamental new constant into physics, the
acceleration of gravity. In fact, neither Galileo nor seventeenth-century scientists in general used the term, "acceleration of gravity." They spoke rather of the distance of body
falls from rest in one second. Quantitatively, this distance
is equal to $\frac{1}{2}\mathbf{g}$; conceptually, it differs in using distance instead of acceleration. Nevertheless, I shall frequently refer
to this distance as the "acceleration of gravity," partly for
convenience, but mostly to insist implicitly that the natural
constant we call the acceleration of gravity was effectively
under consideration. Galileo was convinced that all bodies
fall with the same acceleration, which was then a constant
of nature within his realm of possible experience. He apparently tried to measure it by extrapolating from his experiments on inclined planes. Since the water clock would
not suffice here, he measured a second by means of his pulse.
Moreover, he was unaware that rolling absorbs more than
a quarter of the total energy in generating rotational motion. Perhaps these two factors can explain how he arrived
at a result, that a body falls a bit more than seven feet from
rest in one second, which is only about half the value we
accept.[9]

Meanwhile, Galileo had another potential timepiece, the pendulum. It is not misleading to call him the inventor of the pendulum, for it was Galileo who injected it into science. The pendulum became the most important instrument of seventeenth-century science, and not just as a timepiece. With its inherent capacity of accumulating tiny increments until they become perceptible, it was an instrument capable of yielding great precision. Without it, the seventeenth century could not have begot the world of precision.

As every student of Galileo knows, he gave an incorrect description of pendulums. More than once he asserted that simple pendulums completed all their swings, regardless of amplitude, in the same time. Take two pendulums of the same length. Pull one aside eighty degrees and the other five degrees, and they will swing in perfect synchrony so that no difference between them is perceptible even after a hundred swings.[10] This is, of course, simply not true. After only two swings the difference between the two is perceptible, and they are a full 180° out of phase in less than twenty swings. Galileo's statements about pendulums constitute a major mystery on the part of a man who repeatedly insisted, usually with some asperity, on the ultimate authority of observed phenomena. Perhaps on this topic we should apply to him his own comment about Copernicus believing in the heliocentric system despite the evidence of his senses. With the insight of genius he was able to perceive the potential of the device despite the deficiencies of the pendulums he could observe. It is a fact that by the end of the century, building on the foundation Galileo provided, science had learned to measure time, and with time other things as well, with a precision never dreamed of before.

Galileo himself described the first precision clock in connection with his method of determining longitude at sea. The method made use of the satellites of Jupiter as a clock in the sky telling the same time to every terrestrial observer. But each observer could use it if he knew the local time at the moment of his own observation, for example, the time elapsed from sunset. Galileo described a pendulum made of a bronze sector, of 12 to 15°, from a circle two or three spans in radius (that is, somewhere between one and two feet), with an iron bar through the center to support it (see fig. 3.1). In order to reduce friction to a minimum, he suggested that one should file the lower sides of the bar to knife edges that would rest on bronze supports and rock to and fro as the pendulum swings. He also suggested filing the two sides of the sector to edges to reduce the resistance of the air. To keep the pendulum in motion one would have to push it repeatedly. A wire or bristle on one side was to turn a cog-wheel one notch with each swing, and gearing would run a counting mechanism. One could calibrate the pendulum by the heavens to calculate the time of a single swing.[11] The device would have been neither automatic nor wholly accurate. I am not aware that one was ever constructed. I am willing nevertheless to describe it as the first precision clock.

Meanwhile, isochronism is not the only feature of the pendulum. It also embodies a principle of conservation of mechanical energy. On each swing the bob returns (nearly)

FIGURE 3.1

to the same height from which it began. Although Galileo did not use this feature of the pendulum as a measuring device, he did employ it to illustrate a conclusion that he introduced into his kinematics as an assumption—to wit, that the velocities acquired by a body moving without friction down planes of different inclinations are equal when the vertical displacements are equal. When pendulum AC falls through arc CB, it acquires enough momentum to lift it through arc BD to the level from which it began (see fig. 3.2). Now place a nail at E; the pendulum ascends to G. With the nail at F, the pendulum ascends to I, where G and I are in the horizontal line CD. But the momentum of the body at B, after the descent through arc CB, is equal to that it would acquire in descending through DB, and since the pendulum at B can ascend indifferently through BD, BG, and BI, it must acquire equal quantities of momentum in descending through those arcs. The arcs of course stood here in place of their chords considered as inclined planes, and Galileo used the undefined concept of momentum as a synonym for velocity.[12] Though Galileo utilized this feature of the pendulum solely as an illustration, others would employ

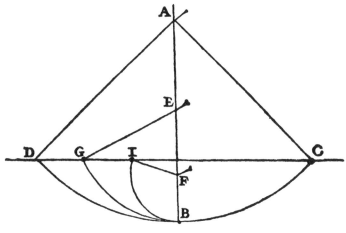

FIGURE 3.2

it as a measuring device, and the pendulum's conservation
of energy would be as important as its isochrony in a physics pursuing precision, as seventeenth-century physics did.

To seventeenth-century science Galileo bequeathed both
the pendulum and a grossly imperfect measurement of **g**.
Those who followed him used the instrument to correct the
measurement. In the 1630s Marin Mersenne convinced himself that a pendulum three and a half feet long swings in
one second (that is, from here to there, half of one complete
oscillation). The foot in use here was the Parisian foot, which
is about one-fourteenth longer than the English foot to which
we are accustomed. Apparently Mersenne used such a pendulum in determining that a body falls 48 feet in two seconds and 108 feet in three, and he calculated from these results that in one second a body falls 12 feet from rest.[13]
Mersenne's measurement, published in the *Harmonie universelle*, implied that a body falls three feet from rest in half
a second—less than the length of the pendulum that swings
through a quadrant of a circle, descending a distance equal
to its length in half a second. This finding stood in direct
contradiction to his analysis of fall in comparison to pen-
dular motion in the same work. In that analysis, he concluded that in half a second a body falling from rest covers
a distance equal to that through which the bob of a seconds
pendulum moves in the same time, that is, the arc of a quadrant of a circle whose radius equals the length of a seconds
pendulum. If the seconds pendulum is three and a half feet
long, a body should then fall very nearly five and half feet
in half a second and about twenty-two in one second.[14]

Later Mersenne devised a different experiment using the
seconds pendulum, which he had now decided was only three
(Parisian) feet in length, to measure **g**. He attached a three
foot pendulum to a wall, held the bob (with the string
stretched perpendicular to the wall) and a ball together, and
released both at the same instant, so that the bob swung
through an arc of ninety degrees into the wall and the ball
dropped onto a platform. By raising or lowering the platform, he found the point from which the two sounds of im-

pact coincided, and his results seemed clearly to indicate that in half a second a body falls more than five feet from rest (see fig. 3.3). Mersenne was unwilling to abandon his earlier measurement, apparently failing to consider that the shorter period of the shorter pendulum made the discrepancy far greater than it appeared to him. A convoluted argument based on his experimental knowledge that a simple pendulum is not isochronous and a swing through a full quadrant consumes more time than a shorter swing convinced him that the time involved in his new measurement was more than half a second, so that the longer distance of fall corresponded to the square of the longer time.[15] Thus he continued to assert that in half a second a body falls three feet from rest, and in one second, twelve. The measurement,

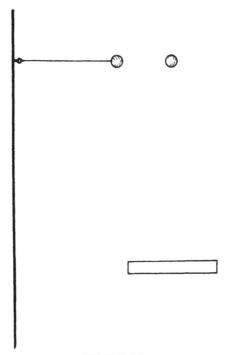

FIGURE 3.3

which is about 80 percent of our value, represented a major improvement on Galileo's.

At much the same time, the Italian Jesuit Giovanni Battista Riccioli determined that the seconds pendulum is nearly three feet four inches long. Riccioli used the Roman foot, which is very close to our foot. Since his goal was precision, he undertook repeated measurements of the pendulum against the heavens, including one with a corps of nine fellow priests who kept the pendulum in motion and counted its swings through a twenty-four hour period until the sun returned to the meridian, to correct the length, and eventually concluded that the exact length is 3 feet, 3.27 inches. Like Mersenne, he then used the seconds pendulum to measure **g**. First, however, he constructed a more refined chronometer. From Galileo's square root rule, corrected by empirical practice, with a shorter device swinging beside the seconds pendulum, he made a pendulum 1.15 inches long that beat six times per second. How could one count something so quick? He called again upon his Jesuit colleagues, and utilizing their musical tradition, plus his knowledge that what musicians called a semichromatic note in standard tempo consumed about one-sixth of a second, he trained them to chant in rhythm with the pendulum. With this arrangement he measured the time of fall from various towers in Bologna. To speak accurately, Riccioli did not understand himself to be measuring **g**, for he was convinced, and his experience bore him out, that all bodies do not fall at the same rate. His standard of reference was an earthen ball (chalk or clay) eight ounces in weight; he concluded that it fell fifteen (Roman) feet in one second, a figure well within 10 percent of our value for **g**.[16]

It remained for Christiaan Huygens to carry the determination of **g** to the highest degree with the realization that any measurement based on the assumption that the simple pendulum is isochronous must be in error. He was able to demonstrate, however, that there is a curve, the cycloid, in which all swings are isochronous (see fig. 3.4).[17] Huygens proceeded then to develop a new mathematical concept, the

evolute, and to demonstrate further that the evolute of a cycloid is an identical cycloid (see fig. 3.5).[18] It followed then
that a pendulum on a flexible line hung between cycloidal
cheeks would move in a cycloid so that all swings would be
truly isochronous. In 1659 Huygens was able to construct
the first truly precise clock (see fig. 3.6). With an escape-
ment, it dispensed with the need for teams of priests to keep
it in motion. It could count the number of swings in a day
measured astronomically, and by the adjustment of its length
with a screw it could be set to swing in exactly one second.
Thus Huygens was able to measure the length of the seconds
pendulum with a precision beyond that of Riccioli.[19]

Initially, Huygens intended to pursue the same avenue
Riccioli had followed and to use the seconds pendulum to
measure **g**. He now realized that he had in fact already done
so, for he had demonstrated the ratio of the time for one
oscillation in a cycloid to the time for a body to fall along
the diameter of the generating circle; in effect, he had a formula for the period of the pendulum in terms of its length
and **g**. Thus he could calculate that in one second a body
falls from rest through approximately 15 (Parisian) feet, 1

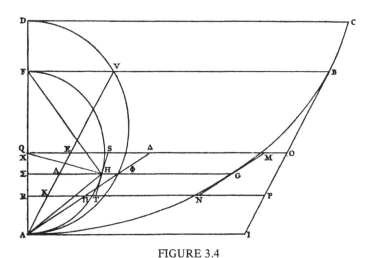

FIGURE 3.4

inch, as he put it in the *Horologium*, or more exactly, 15 feet, 1.1207 inches, in effect the figure we continue to accept for g.[20] Near the middle of the seventeenth century then, by means of the pendulum, scientists had precisely determined the value of a physical constant that had been known half a century earlier and beyond measure when it was defined, and they had learned to measure a quantity, time, earlier measurable only in gross terms, with a degree of precision that exceeded any other measure.

All of this formed the background to the career of Isaac Newton, who matriculated in Cambridge shortly after Huygens perfected the precision clock, through before he published the theory that supported his invention. Newton assimilated, as though by instinct, the new spirit of *quantitative precision*. A quarter of a century later, in the *Principia*, he raised the tradition of mathematical science to a new level, defining a quantitative dynamics to explain both Galileo's and Kepler's quantitative kinematics, and thereby establishing once and for all the character of modern science. Newton also understood the necessity of a quantita-

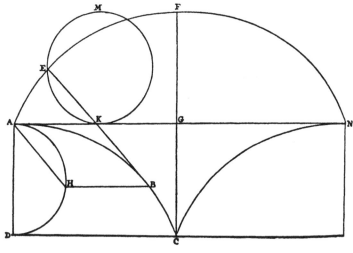

FIGURE 3.5

tive terrestrial physics for a science that claimed to give an exact account of empirical reality. And more than anyone else, he comprehended the rich potential of the pendulum.

The pendulum figured in Newton's physics almost from the beginning. Among his earliest manuscripts is one—known as the Vellum Manuscript because it is written on the back of a lease for a piece of property his mother owned—in which Newton undertook to solve a problem that he found in

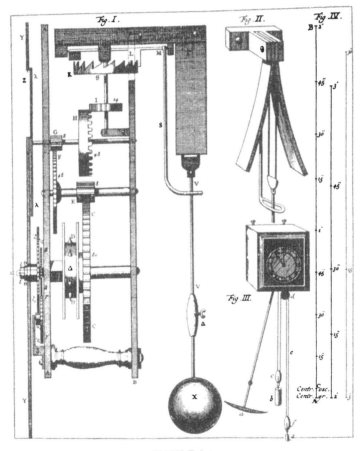

FIGURE 3.6

Galileo's *Dialogue*.[21] One of the arguments against the motion of the earth was the assertion that a diurnal rotation would fling loose objects off into space. Galileo's answer to the objection contained serious flaws. The question caught Newton's attention sometime near the beginning of 1665, a few years after Huygens's analysis of circular motion but before its publication in the *Horologium*. Newton's quantitative demonstration that the measured acceleration of gravity is many times greater than any centrifugal effect caused by the earth's motion is a perfect expression of the new spirit that was conquering natural philosophy.

After a false start based on the inadequate understanding of the forces associated with circular motion, Newton undertook an analysis of such motion that arrived at a mathematical expression in terms of velocity and radius identical to our formula for centripetal force.[22] In 1665, however, the expression represented the tendency of a body moving in a circle to recede from the center, what Huygens called centrifugal force, for it was in those terms that Newton then conceived of the dynamics of circular motion. On the Vellum Manuscript, he proceeded to calculate from the size of the earth and its rate of rotation that the force of gravity is 150 times greater than the centrifugal force.

But his measure of gravity came from Galileo, and before Newton set the Vellum Manuscript aside, he realized that he could check the figure via a pendulum. His device was a conical pendulum. As Newton conceived of it at that time, the bob of a conical pendulum is held in equilibrium by three forces, its gravity, its centrifugal force, and the tension in the string. Rearrange the diagram of forces in a triangle, the two acute angles of which are set by the inclination of the pendulum. The proportion of the vertical leg of the triangle to the horizontal leg gives the proportion of gravity to centrifugal force (see fig. 3.7). According to Newton's new analysis, the centrifugal force of a body moving uniformly in a circle is equal to the force that will generate the same quantity of motion in the body if applied uniformly during the time it takes the body to move one radian, or (which is the same thing) equal to the force that will move an equal body from rest through half the radius of the circle during the same time. It is easy to show that this conclusion is quantitatively identical to the formula we apply to circular motion today. The second version of the expression, together with the measure of time by movement through one radian, implied that when bodies move in concentric circles, if the centrifugal forces are proportional to the radii of the circles, all circles will be completed in equal times. Recall the force diagram of the conical pendulum. In all conical pendulums of the same height, the centrifugal forces are proportional to the radii of the circles (see fig. 3.8). It follows that all conical pendulums of the same height have the same period.[23]

A conical pendulum moving uniformly in its circle is an

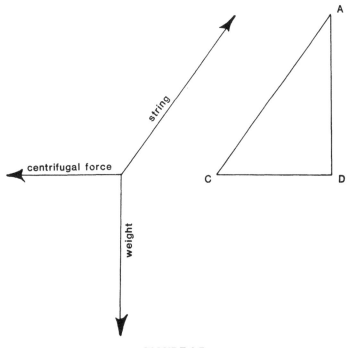

FIGURE 3.7

ideal fiction. However, the pendulum of the same height that traces an infinitesimal circle will have the same period, and the pendulum will be indistinguishable from a simple pendulum of the same length swinging through an infinitesimal arc. Thus there was the possibility of measuring the period of a simple pendulum—with small arcs it would he sufficiently isochronous—and then converting it in thought to a theoretical conical pendulum of equal height. Any conical pendulum would do, but a 45° pendulum, with centrifugal force equal to gravity, would yield the simplest calculations. The Vellum Manuscript carried out its calculations in terms of 45° pendulums. It is a fascinating document. The first calculation, which contains a combination of errors, arrived at the conclusion that Galileo's measure of **g**, far from being too small, was three times too large. Newton found the worst error, and his second calculation indicated that Galileo's figure was roughly correct. But he kept working on the problem, finding and correcting his errors and deepening his understanding.

To measure the period of a pendulum he had to have some

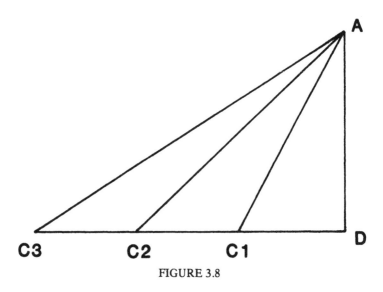

FIGURE 3.8

measure of time. The manuscript contains two related, peculiar ratios.

$7/54 = 56/x$ (with the computation that x = 432)

$6/21 = 432/y$ (with the computation that y = 1512)

1512, the manuscript informs us, was the number of oscillations in one hour. That is, 7 units of some sort corresponded to one-eighth of an hour. These 7 units taken together contained 54 smaller units. A pendulum completed 21 swings in 6 of the smaller units and thus would have complete 1512 in an hour. I am convinced that Newton's chronometer was the sun.

As the cases of Mersenne, Riccioli, and Huygens indicate, before the precision clock appeared, only the heavens offered an exact measure of time. A number of stories establish that Newton was an accomplished sun-dialer. I am inclined to think that he used a pinhole image, perhaps with a lens to sharpen the edges. I am willing to speculate that he had experimented in the precise measurement of an eighth of an hour and had calculated the distance at which the image of the sun would cross seven of some arbitrarily given units that he found at hand, and that when he wanted to measure smaller units he divided the arbitrarily given ones into fractions of an inch. Whether or not my speculations are correct, it is a fact that the Vellum Manuscript records the success of a student without any sophisticated equipment in using small variations of the ratios above to measure the periods of two pendulums to about one percent of the correct values. From the periods he calculated the distance that a body would fall from rest in one second, arriving at a value again about one percent lower than ours. When Galileo's value of **g** turned out to be only about half the correct value, Newton did not repeat the original calculation about gravity and centrifugal force on a rotating earth but simply doubled his answer. The force of gravity is three hundred times larger. The Vellum Manuscript offers a clear prognostication of what might be expected from the young man who composed it.

About fifteen years after the calculations on the Vellum Manuscript, Newton used a pendulum to test whether an aether exists. According to the prevailing mechanical philosophy of nature, which Newton had met and adopted during his undergraduate years, the physical world consists solely of bodies in motion, which can affect each other only by direct contact. Thus mechanical philosophers explained the solar system in terms of a huge vortex of matter that carried planets around the sun much as a river carries branches and other bodies that have fallen into it. The matter that composed the vortex was usually called the aether because of the subtle size of its particles; mechanical philosophers called upon the aether, which was understood to be imperceptible to the senses, to explain phenomena that did not depend on the impact of any manifest particles. Thus they used the pressure of the aether to explain gravity—that is, the tendency of bodies near the surface of the earth to descend. It had not taken Newton long after he embraced the mechanical philosophy to begin to have second thoughts about it. Some of the second thoughts derived from religious considerations, for Newton rejected the notion of the physical world as an autonomous realm governed solely by the laws of mechanical necessity. But some of the second thoughts had to do with natural philosophy. He began to entertain the notion that various phenomena, such as capillary action and chemical reactions, are caused, not by invisible mechanisms in an imperceptible aether, but by forces of attraction and repulsion that act at a distance.

He began as well to doubt that an aether, the *sine qua non* of a viable mechanical philosophy, existed. To test whether or not there is an aether, Newton contrived an experiment that utilized the conservation feature of the pendulum. An empty wooden box functioned as the bob on a pendulum eleven feet long. Newton pulled the box six feet to one side and carefully marked the points to which it returned on its first four swings. Then he filled the box with metal so that it weighed 78 times as much. Exercising great care, he readjusted the length of the pendulum's string. In each case he added half the weight of the string to the weight of the box, and he even calculated the weight of the air enclosed in the empty box. If gravity, or heaviness, is produced by the pressure of an aether, it must penetrate bodies and strike against all of their particles, for the weights of bodies of the same substance increase in proportion to volume, not to surface. An aether able to penetrate bodies in this manner would have also to resist motion through it. Indeed, in writings before this experiment, Newton always cited the fact that pendulums in the evacuated receiver of an air pump come to rest as quickly as they do in the air as evidence for the existence of an aether. In his specially contrived experiment, however, the heavier pendulum required nearly 78 swings to decay to the mark reached by the empty box on its first swing, nearly 78 more to reach the second mark, and so on. In fact, it required 77 swings; Newton was convinced that the slight deficiency from the number 78 was due, not to the drag of an aether, but to the greater resistance of the air to the heavier pendulum, which moved longer at a higher velocity than the lighter one.[24] The experiment, which apparently dated from about 1679, convinced Newton that there is no aether. Thus it played a major role in clearing the way for the *Principia*.

Newton began to work on what became the *Principia* late in 1684, as a result of a visit from Edmond Halley. Almost his first step was to clarify basic concepts of dynamics, one of which, quantity of matter, or mass, entered the vocabulary of physics at this time.[25] How can one measure the mass of a body? It is usually proportional to the body's weight, Newton replied, but he went on to suggest a device by which it could be measured independently of weight. Suspend equal weights on pendulums of equal lengths; the masses of the two bodies will be inversely proportional to the number of oscillations they complete in the same time.[26] On the manuscript, Newton crossed this statement out, and in an empty space opposite it he recorded the following note: "When experiments were carefully made with gold, silver, lead, glass, sand, common salt, water, wood, and wheat, however, they resulted always in the same number of oscillations."[27] The precise proportionality of mass to weight—a necessary consequence of the concept of universal gravitation in Newton's formulation of it, in itself an embodiment of the concept—was another gift that the pendulum offered to modern physics.

In the fully elaborated argument of the *Principia*, no measured quantity plays a larger role than **g**. Newton chose to rely here, not on his own early measurement, but on Huygens's, which was more accurate and was also known and accepted by the scientific community. From Kepler's third law and from the quiescence of the planets' lines of apsides, Newton could establish that the sun must attract the planets with a force that varies inversely as the square of the distance; and since the satellites of Jupiter also obey Kepler's third law, he could show that Jupiter must attract them with a similar force. Furthermore, the fact that the moon circles the earth indicates that the earth must attract it, but only the correlation of the moon's orbit with **g**, according to the same inverse square relation, allowed Newton to connect celestial phenomena with terrestrial ones and to apply the ancient word *gravitas*, heaviness, to the cosmic force.[28] The persuasiveness of the argument rested directly on the quantitative precision of the correlation. It is not too much to assert that without the pendulum there would have been no *Principia*.

The measurement of **g** was by no means the only use of the pendulum in Newton's great work. The Corollaries to the Law of Motion offered a principle, equivalent to the conservation of momentum, to determine the motions of bodies in impact. Was it possible to test the principle to demonstrate its validity? Newton called upon experiments with pendulums that were somewhat analogous to Galileo's illustration of his assumption that bodies acquire the same increment of velocity for equal vertical descents regardless of the paths they follow. In Newton's experiments the bodies, suspended from equal pendulums, were released at the same instant and met in impact at the lowest points of their trajectories. From the heights of fall Newton could compute their velocities at impact, and from the heights to which they rose he could compute their velocities after impact.[29] He was not only able to confirm the conservation of momentum in impact, but also introduced a new quantity into physics, what he called "elastic force" and what we know today as coefficient of elasticity. His specific conclusions about different materials could not have been much less incredible to his age than the principle of universal gravitation—to wit, that glass is nearly twice as elastic as steel.[30]

Book II of the *Principia* developed a theory about the resistance of material media to motion. He tested the theory by means of experiments with long pendulums that were similar to his test for the existence of a aether a few years earlier. He let pendulums oscillate in the air, in water, and (with a contrivance he worked out) even in mercury in order to measure the resistance offered by materials of greatly different densities.[31]

To establish the shape of the earth, Newton called upon the experience that near the equator a pendulum had to be shortened from its European length in order to oscillate in one second.[32] That is, once **g** had been measured by a pendulum in Europe, differences in the length of pendulums could measure small variations in **g** in other latitudes. Newton himself never traveled beyond the borders of England, and in this case he called upon the measurements of others, primarily members of French scientific expeditions, which he was able to fit into a pattern of comprehension.

The *Principia* even used the pendulum in its basic capacity, as a chronometer, in the measurement of the speed of sound. As part of Book II, Newton defined the dynamic elements in the propagation of sound and calculated its velocity through air. When he turned, as he always did, to compare theory with empirical reality, he found that there was no established measurement of the speed of sound. Measurements had varied wildly from 600 to 1500 feet per second, hardly the degree of precision to which Newton aspired. In a portico 208 feet long, with a good echo, in Trinity

College, Newton adjusted a pendulum to swing to and fro with the return of the first and second echoes. He found that the pendulum had to be longer than $5^1/_2$ inches and shorter than 8 inches, corresponding to a speed of sound between 9 and 1035 feet per second. His calculated speed fell squarely between the two.[33] With the second edition, speed of sound became a problem because new measurements, which Newton trusted, had placed it above his upper limit, and he introduced a number of ad hoc considerations into his calculation in order to increase the result.[34] If the new measurements, which arrived at substantially the figure we still accept, were above his original upper limit, however, they were not far above it. By an imaginative application of the pendulum Newton had effected a considerable improvement over previous notions of the speed of sound.

Newton's *Principia* appeared in 1687, only a couple of years less than a century after Galileo composed the notes on *De caelo* that I cited earlier. During that century, scientific thought had moved from one world into another, and the central characteristics of the new world were mathematics and quantitative precision. Theory was now acceptable only if it corresponded, not to the text of Aristotle, not to the word of Scripture, not to plausible arguments, not to observed qualities, but to the quantities found in nature. Largely without the elaborate instruments that we now assume as integral parts of the scientific enterprise, though not without one simple one, science had learned during the century to measure nature. It appears to me that there has not been a more fundamental change in the history of Western thought.

NOTES

1. *Le opere di Galileo Galilei*, ed. Antonio Favaro, 20 vols. in 21, Firenze, 1890–1901, 1:1–177.

2. William Wallace, *Galileo and his Sources. The Heritage of the Collegio Romano in Galileo's Science*, Princeton: Princeton University Press. 1984. Adriano Carugo and Alistair C. Cromble, "The Jesuits and Galileo's Ideas of Science and of Nature," *An-*

nali dell'Istituto e Museo della Scienza di Firenze 8.2 (1983):3–68.

3. Colombe, *Discorso apologetico*; *Opere* 4:328, 336–37, 340–41, 347.

4. *Discoveries and Opinions of Galileo*, Stillman Drake, trans. and ed. (Garden City, N.Y.: Doubleday. 1957), pp. 237–38.

5. Johannes Kepler, *Harmonice mundi*, Bk. IV, Chap. 1; *Gesammelte Werke*, ed. Max Caspar, 19 vols. (Münich, 1938–75), 6:223.

6. *Risposta alle opposizioni*; *Opere*, 4:563 and 573–75. Also see the extended discussion of resistance to division, pp. 515–83.

7. Galileo, *Two New Sciences*, Stillman Drake, trans. (Madison: University of Wisconsin Press. 1974), pp. 169–70.

8. Thomas B. Settle, "An Experiment in the History of Science," *Science*, 133 (1961):19–23.

9. David Lindberg, "Galileo's Experiment on Falling Bodies," *Isis*, 56 (1965):352–54.

10. *Dialogue Concerning the Two Chief World Sytems*, tr. Stillman Drake (Berkeley: University of California Press. 1962), pp. 230 and 450. *Two New Sciences*, pp. 87–88 and 97. Galileo to Real, June 1637; *Opere*, 17:100–101.

11. Galileo to Real, June 1637; *Opere*, 17:101–102.

12. *Two New Sciences*, pp. 162–64.

13. M. Mersenne, *Harmonie universelle* (Paris, 1636), Bk. II, "Des mouvements de toutes sortes de corps," pp. 87 and 135–36. See Alexander Koyré, "An Experiment in Measurement," in his volume of collected essays on the scientific revolution, *Metaphysics and Measurement* (Cambridge: Harvard University Press. 1968), pp. 89–117.

14. *Ibid.*, p. 132.

15. M. Mersenne, *Novarum observationum physico–mathematicarum tomus III* (Paris, 1647), pp. 152–59.

16. Giovanni Battista Riccioli, *Almagestum novum* (Bologna, 1651), vol. I (part 1), bk. II, chaps. XX–XXI, pp. 86–90; vol. I (part 2), bk. IX., sect. IV, chap. XVI, pp. 384–89. See Koyré, "An Experiment in Measurement."

17. Christiaan Huygens, *Horologium oscillatorium* (Paris, 1673), pp. 57–58.

18. *Ibid*, p. 67.

19. 3 (Parisian) feet, 0.708 inches (or, as Huygens gives it, 3 × 881/864 feet). *Ibid.*, p. 155.

20. *Ibid.*, p. 155.

21. Cambridge University Library, *Add. MS. 3958*, f. 45. It is published, together with a photographic reproduction and an analysis of its contents, both in *The Correspondence of Isaac Newton*, ed. H. W. Turnbull, et al., 7 vols. (Cambridge: Cambridge University Press. 1959–77), 3:46–54, and in John Herivel, *Background to Newton's Principia* (Oxford: Oxford University Press. 1965), pp. 183–91.

22. Cambridge University Library, *Add. MS. 4004*, f. 1; published in Herivel, *Background*, pp. 129–32.

23. *Ibid.*, p. 131.

24. Newton described the experiment in detail in the *Principia. Sir Isaac Newton's Mathematical Principles of Natural Philosophy and his System of the World*, tr. Andrew Motte, ed. Florian Cajori (Berkeley: University of California Press. 1934), pp. 325–26.) The English edition, which derives from the third Latin edition, does not contain a final sentence, found in the first edition, which traced the deviation from 78 swings to the resistance of the air. *Isaac Newton's Philosophiae naturalis principia mathematica*, the third edition (1726) with variant readings, ed. Alexander Koyré and I. Bernard Cohen, 2 vols. (Cambridge: Cambridge University Press. 1972), 1:463.) In his old age, Newton was willing once more to admit the existence of an aether, though an aether very different in its rarity from the standard aether of the mechanical philosophy (as the description of the experiment, which he continued to include in the *Principia*, insisted); with the second edition of 1713, he removed the final sentence about the resistance of the air.

25. The manuscript is published in Herivel, *Background*, pp. 306 and 315.

26. *Ibid.*, p. 306.

27. *Ibid.*, p. 317. Herivel's publication gives a mistaken impression of where this passage appears in the manuscript.

28. *Principia*, pp. 407–408, 482–83.

29. *Ibid.*, pp. 22–24.

30. *Ibid.*, p. 25.

31. *Ibid.*, pp. 316–25. The second and third editions offered further experiments of bodies falling through water and air (pp. 355–66).

32. *Ibid.*, pp. 430–33

33. I cite from the first edition; Koyré-Cohen, *Principia*, 1:30.

34. *Principia*, pp. 382–84. I have discussed this in detail in "Newton and the Fudge Factor," *Science*, 179 (1973):751–58.

4

Newton's 'Mathematical Way' a Century After the *Principia*

Thomas L. Hankins

Any attempt to evaluate Newton's legacy a century after the *Principia* requires that we ask the question, "What was different about physics in 1787? Was it "Newtonian" in the sense that it followed the mathematical techniques and the experimental methodology of Newton, or had it changed in any significant way? Had Newton's method finally triumphed over those of Descartes and Leibniz, as Voltaire proudly claimed, and established itself as the proper method for all the physical sciences? Or had it been replaced by something different? Or, putting it a different way, were natural philosophers still going about their business in the same way that Newton did?

At first sight we would have to conclude that there had been significant changes. The most striking event marking the first centenary of Newton's *Principia* was the publication of Lagrange's *Mécanique analytique*. Lagrange's work does not look at all like Newton's. Lagrange boasted that his book contained no figures. In fact his boast actually went further than that.

He said: "One will not find any figures in this work. The methods that I present require neither constructions, nor geometrical or mechanical reasoning, but only algebraic operations subjected to a regular and uniform progress. Those who love analysis will see with pleasure mechanics become a new branch of that science and will thank me for having extended thus its domain."[1]

Lagrange's work is, indeed, entirely analytical in stark contrast to the *Principia* which is entirely geometrical. This transformation from a geometrical to an analytical formulation of mechanics had been going on throughout the century. Jean d'Alembert, in his article "Analyse" for the great French *Encyclopédie*, credited Charles Reyneau with being the first to introduce analytical methods into mechanics in his *Analyse demontrée* of 1714. Clifford Truesdell more recently has claimed Leonhard Euler's *Mechanica* of 1736 as the first real treatise on analytical mechanics. All would agree, however, that Lagrange's *Mécanique analytique* was the culmination of a trend that had transformed mechanics.

Not only does the *Mécanique analytique* look different from the *Principia* because it is full of algebra instead of geometry, but also its methods are different. Lagrange uses variational methods employing d'Alembert's principle, the conservation of vis viva, and the principle of least action, all principles that differ substantially from Newton's laws. In fact Lagrange, while showing his great respect for Newton, repeats the criticisms of the concept of force that were common in France throughout the eighteenth century. All three of Newton's laws of motion are based on the concept of force and describe the motions of point masses. As a protege of d'Alembert, Lagrange shared d'Alembert's desire to build the science of mechanics without using forces and to extend it to constrained systems of masses, rigid bodies, and continuous media. Analytical methods helped the continental mathematicians achieve those goals, but it carried them far from Newton's original formulation of mechanics.

If we turn now to experimental physics we see that it also had changed greatly by 1787. Historians of science have frequently noted that experiment during the 1780s became much more quantitative, apparatus became more precise, and experimenters concentrated on getting numbers rather than wondrous effects. The two examples most commonly cited as evidence are Charles-Augustin Coulomb's torsion balance, with which he measured the force of electrical attraction in 1785, and the precise experiments on the weight of heat by Benjamin Thompson (Count Rumford) in 1787.

There are other examples of quantitative experimentation, however, that are more interesting even though they may not be quite so close to the centenary date of 1787 that we have been considering. In 1781 Alessandro Volta, who would invent the voltaic pile, came to Paris where he carried out a series of experiments with the mathematician Pierre-Simon Laplace and the chemist Antoine-Laurent Lavoisier. Lavoisier and Laplace had joined forces in 1777 to do experiments on evaporative cooling.[2]

Lavoisier already had his oxygen theory of combustion well in hand, but he had not yet dared attack the phlogiston theory, because he did not understand what was happening to the heat in processes such as combustion and evaporation. He recognized Laplace's analytical skills and persuaded Laplace to join him.

Laplace was somewhat reluctant to leave mathematics for experimental physics, but he saw the advantages of collaborating with a distinguished scientist like Lavoisier, and as he explained to his friend Lagrange: "Because he [Lavoisier] is very rich, he does not spare any expense in order to give our experiments the precision that is necessary in such delicate researches."[3]

Precision was a new goal of experimenters and it was expensive. Lavoisier's wealth was an important adjunct to his skill as an experimenter.

In order to measure heat precisely, Laplace designed and Lavoisier paid for the construction of two ice calorimeters, which measured the amount of heat produced by the weight of ice melted. The experiment most commonly cited is Lavoisier and Laplace's measurement of the heat produced by a guinea pig, but in 1782, a more important problem was measuring the specific heats of metals and their calxes (oxides), because if the metal had a greater capacity for heat than its oxide, then it would appear that the caloric, or matter of heat, or matter of fire, whatever one called it, was released by the metal as phlogiston. Fortunately the ice calorimeter showed that the calx, or oxide, had a greater specific heat, which gave Lavoisier a powerful weapon against the phlogiston theory. Such results were obtainable only by precise calorimetric measurement.[4]

Volta arrived in Paris in December 1781 as part of a grand tour of the scientific centers of Europe. He brought with him his new condensing electroscope, which allowed him to detect and measure extremely small electrical charges. Lavoisier and Laplace, who had completed their original experiments on evaporation, saw that Volta's very sensitive electroscope might allow them to measure an electrical charge produced by evaporation. They hoped to detect the source of atmospheric electricity (and perhaps the cause of lightning). With the condensing electroscope they were able to detect an effect. This indicated that perhaps the substance of electricity and the substance of heat were involved in similar ways in the process of evaporation.[5] All of these conjectures on heat and electricity required new instruments and precision measurement, because the effects were small and their magnitudes had to be compared.

Another sign of the new concern for quantitative experiment in the 1780s was an interest in error calculation. Earlier eighteenth-century experimental physicists had not attempted to judge their errors and had shown little or no concern for significant figures. Usually the number of significant figures in a result was a measure of the author's desire to impress his reader rather than a measure of his accuracy.[6]

Laplace had become interested in error calculation in 1772 but it was not until 1777, the year that he joined forces with

Lavoisier, that error calculation became a subject of discussion at the Paris Academy of Sciences. John Heilbron, in his book *Electricity in the 17th and 18th Centuries*, says that Henry Cavendish was probably the first person to attempt a real error calculation, when, in 1771, he estimated the error in his measurement of the force of electrical attraction.[7] The beginning of an interest in error calculation is one of the best bits of evidence that the attempt to make experiment more quantitative was conscious, and not just a historical accident. One doesn't worry about measuring errors when one isn't interested in quantitative effects.

As I have indicated above, rational mechanics and experimental physics both showed marked changes in the 1780s. Mechanics finally became totally analytical while experimental physics became more quantitative and its instruments more precise. We now need to ask: 1. "Were these two kinds of changes in physics related or were they independent?" and 2. "Were they a departure from Newton's method or were they a continuation of the Newtonian tradition?" In answering the first question, historians of science have tended to argue that they were independent. I. B. Cohen has argued that Newton's *Principia* and his *Opticks* were the foundations of two very separate disciplines in the eighteenth century—rational mechanics and experimental physics.[8] Thomas Kuhn in a widely read article entitled "Mathematical versus Experimental Traditions in the Development of Physical Science" makes a clear dichotomy between the mathematical and the experimental approaches to physics.[9] Clifford Truesdell, in an equally well-known article on the history of rational mechanics, argues that rational mechanics in the eighteenth century was entirely mathematical and independent of practical mechanics or physics.[10] Granted that the two traditions of mathematics and experiment merged in the early nineteenth century, can we accept their conclusion that the two traditions were still separate in the 1780s? I would argue that the scientists themselves never thought they were separate, and that the moves toward analytical methods in mechanics and quantitative methods in experiment had a common source in what the eighteenth century called "analysis."

The answer to the second question, whether the changes that we have noted in mechanics and experimental physics constitute a departure from Newton's method, also turns on what was meant by "analysis," because both Newton and his successors in the eighteenth century regularly used this term to describe their procedures.

Unfortunately we use the term "analysis" very loosely, and so did the philosophers of the seventeenth and eighteenth centuries. In fact the problem goes back to antiquity, where Aristotle established the word in his Prior and Posterior Analytics and where Pappus established a comparable method in geometry.[11] But modern scholars have not been able to agree over the precise meaning of Pappus's remarks. Some argue that by geometrical analysis Pappus meant moving back up the chain of implications from an assumed conclusion to premises that are known to be true.[12] Others argue that he meant proceeding deductively from an assumed conclusion until one reached a proposition known to be true and then attempting to reverse the process.[13] Still others argue geometrical analysis depends on the crucial role of auxiliary construction.[14] Some even argue that the confusion in Pappus's description of his method arises from the fact that the passage in question was not all by Pappus![15] A later interpolator may have sneaked in an "explanatory" sentence that in fact makes the passage ambiguous. Because there are almost no other texts against which to compare Pappus's remarks (the most important is an interpolated passage in Euclid), the debate is not likely to be resolved. It also suggests the possibility that even in antiquity, the term "analysis" lacked a single unambiguous meaning.

In the Latin West analysis and synthesis were more commonly called the methods of Resolution and Composition. Resolution meant moving from observed particulars to general principles. Composition meant moving the other way from general principles to the explanation of particular events. In the seventeenth century analysis and synthesis were usually regarded as two separate methods. The Port Royale *Logic* of 1662 argued that analysis was the method of invention, useful for discovering truths, while synthesis was a method of demonstration or of instruction for leading others to the truth. The Port Royale *Logic* regarded the two methods as being quite distinct. Galileo typifies another seventeenth century interpretation of these two methods. He tended to think of analysis as conjectural, while synthesis was a method of demonstration. Analysis or resolution was, for Galileo, only a suggestive procedure for propositions that would then be demonstrated or tested by mathematical proof or by experiment.[16]

Newton's statements about analysis indicate that he, also, thought of analysis as a general method applicable to both mathematics and experiment. In the *Opticks* he wrote:

> As in Mathematics, so in Natural Philosophy, the Investigation of difficult Things by the Method of Analysis, ought ever to precede the Method of Composition [or Synthesis]. This Analysis consists in making Experiments and Observations, and in drawing general Conclusions from them by Induction, and admitting of no Objections against the conclusions, but such as are taken from Experiments, or other certain Truths. For Hypotheses are not to be regarded in experimental philosophy. . . . By this way of Analysis we may proceed from Compounds to Ingredients, and from Motions to the Forces producing them; and in general from Effects to their Causes, and from Particular Causes to more general ones, till the Argument end in the most general. This is the Method of Analysis: and the Synthesis consists in assuming the Causes discovered, and established as Principles, and by them explaining the Phaenomena proceeding from them, and proving the Explanations.[17]

Most significant for our purposes is the first sentence of this famous quotation where Newton states that in natural philosophy, *as in Mathematics*, the Investigation of difficult things by the Method of Analysis, ought ever to precede the Method of synthesis. The method of analysis applies equally well to mathematics, chemistry, and experimental physics. Newton, in this passage, was following his teacher Isaac Barrow, who had stated in his *Mathematical Lectures* of 1664–1665 that "*Analysis* . . . seems to belong no more to *Mathematics* than to *Physics*, *Ethics* or any other Science. For this is only . . . a certain Manner of using Reason in the Solution of Questions, and the Invention or Probation of conclusions, which is often made use of in all other Sciences."[18] Analysis was, then, a very general method to be employed in all kinds of reasoning.

Newton's example of analysis and synthesis in chemistry is the easiest case to understand. Compounds are analyzed to determine their elemental constituents, and the elements can then be brought together in different ways to synthesize new compounds. But the case is not so clear in mathematics. Newton explains the distinction in his *Universal Arithmetick*, which he subtitled "A Treatise of Arithmetical Composition and Resolution." He says that arithmetic is synthetic, because it deals only with known quantities in working toward the unknown answer. Algebra is analytic, because it assumes the unknown answer "as if it were given."[19] In this mathematical case Newton is thinking less about taking things apart, as in the example of chemistry, and more about assuming a conclusion to be true, as in geometrical analysis.

Henry Guerlac has argued that Newton's achievement was to make analysis and synthesis two parts of the same method rather than two different methods.[20] Analysis, for Newton, was not merely a method of conjecture, but an essential part of the process of demonstration in natural philosophy. In the passage quoted earlier from the *Opticks* Newton stated that analysis must always precede synthesis in natural philosophy, because if one does not begin with analysis of the phenomena through experiments, then one must "feign hypotheses" in order to proceed with the synthetic part of the demonstration. Newton's statements about analysis were the basis for most other definitions of analysis throughout the eighteenth century, but one would have to admit that Newton failed to give a precise and unambiguous definition of the term. And to make matters more confusing, his contemporary Robert Hooke used the terms analysis and synthesis in a way exactly opposite to that of Newton. For Hooke, synthesis was a passage from observations to more general propositions and analysis a passage in the opposite direction.[21] In this case the confusion is not a result of applying the terms to different subject matter, but a complete reversal in the meaning of the words.

However confusing the legacy of Newton may have been, the method of analysis and synthesis was associated with him right through the eighteenth century from Voltaire to Laplace. When Voltaire popularized Newton in France in the 1730s, he constantly emphasized the method of analysis and attributed it to Newton. "That great man," said Voltaire, referring to Newton, had always proceeded by analysis and had stopped "whenever this torch was lacking to him."[22] In fact, Voltaire insisted, "the only way man can reason on the objects of experience is by analysis."

Voltaire's views coincided with those of the most important French philosopher to write on the method of analysis, the Abbé Etienne Bonnot de Condillac. Condillac carried the emphasis on analysis even further than Voltaire, so far, in fact that he denied the validity of synthesis altogether. His friend, the mathematician Jean d'Alembert, had echoed Newton's statement about analysis and synthesis in the famous French *Encyclopédie* where he wrote: "In natural philosophy, as well as in mathematics, it is necessary first to clear away the difficulties by the analytical method, before employing the synthetic method."[23] Condillac agreed that analysis should be used in the experimental as well as the mathematical sciences. "I well know," he said, "that it is customary to distinguish different kinds of analysis: logical analysis, philosophical analysis, mathematical analysis; but there is only one kind; and it is the same in all the sciences."[24] But Condillac then went on to claim that analysis was the *only* acceptable method in science. He recognized a double method of "decomposition and composition," but this double method he called analysis. Synthesis, in mathematics as well as in experiment, he believed was only a hidden use of analysis and therefore he denied the validity of synthesis altogether as a separate method of demonstration.

Condillac believed that analysis was the only correct method of reasoning, because it was taught to us by nature herself.[25] "To analyze an object is merely to observe its qualities in successive order so as to arrange them in the mind as they actually exist. . . . The method of analysis, which is believed to be known only by philosophers, is, in fact, known to everyone."[26] According to this egalitarian conclusion, the only way to demonstrate a proposition in physics or mathematics is to go over the actual path of discovery leading to the truth of the proposition, because the path of discovery can only be a process of analysis.

A second important aspect of Condillac's advocacy of analysis was his emphasis on language. In creating a language we associate signs with the objects of knowledge in order to communicate with one another, but more importantly, we use the signs in the process of reasoning. Without signs, that is words, we cannot reason at all. We need a word to stand for each idea, abstract or concrete, that we employ in scientific reasoning. Language is, for Condillac, an analytical system. If you will forgive the anachronism, we might say that according to Condillac, language is the software; the mind is the hardware. The mind can do nothing without words.

In 1780, the last year of his life, Condillac wrote: "The creation of a science is nothing else than the establishment of a language, and to study a science is to do nothing else than to learn a well-made language."[27] If we accept Condillac's argument for the primacy of language in analysis, then the excellence of our analysis will depend on the language that we employ. And this is exactly what Condillac says: "languages include all means that we can have for analysis—good or bad."[28] Language is analysis, and good language is good analysis; bad language is bad analysis. Condillac claimed that algebra was the very best language of all, because it had the best signs. There was no ambiguity in their meaning, and the analytical operations performed with these signs were perfectly clear.

If algebra were the perfect analytical language, as Condillac indicated, then it is, perhaps, not surprising that Lagrange believed his mechanics would be pleasing to those who love analysis. He was certainly thinking primarily of mathematical analysis when he made this statement, but we should realize that the transition from geometrical to analytical mechanics fulfilled and also typified a much broader move to analysis in all parts of natural science, not just in mathematics and rational mechanics.

We should also realize that the method of analysis had not lost its ambiguity. Condillac's analysis was both a grammar and a method of investigation the two parts of which did not always fit together very well. Lagrange confused matters even further for the mathematicians by using the word "analysis" in a variety of ways. Where analysis had once been a method employed in geometry, it now became directly opposed to geometry. Lagrange was also responsible for the term "analytic function" referring to a function that has derivatives of all orders. And to make matters even worse, Gaspard Monge used the term "analytic geometry," to describe the invention of Descartes and Fermat in the seventeenth century. Descartes had called his method "the application of algebra to geometry." The new name "analytical geometry" further confused the meaning of analysis.

Immanuel Kant attempted to clear up this confusion for philosophers by carefully redefining the terms "analysis" and "analytical," but he did it in a way peculiar to his own philosophy, and in two very different senses. For Kant an analytical judgment or proposition was something quite different from an analytical method, which often employs synthetical propositions. Without attempting to unravel Kant's terminological confusion, we note that he, himself, wrote: "It is unavoidable that, as knowledge advances, certain expressions which have become classical after having been used since the infancy of science will be found inadequate and unsuitable, and a newer and more appropriate application of the terms will give rise to confusion."[29] Kant's new definitions aided those reading his books, but did not clarify the meaning of "analysis" as it was commonly used.

Considering the confusion that reigned over the meaning of analysis, it is surprising that the method was so loudly acclaimed in the eighteenth century. It is not unusual, however, for an ambiguous term to be highly useful in science, especially when it expresses a sentiment that resists precise description. Condillac, in his attempt to demonstrate that analysis was the "natural" way to order the world, reinforced the widely held conviction that good method and common sense were one and the same.

If the philosophers of the eighteenth century identified language with analysis, and if they argued that analysis was essential in experiment as well as in mathematics, then one should expect to find language playing an important part in experiment. We can accept Condillac's argument that mathematics already had a good language in algebra. What about experiment? A few such languages did exist, but it is surprising to find that they were more common in natural history than in the physical sciences. Probably the most famous analytical language was the binomial nomenclature for plants proposed by Carl Linnaeus in his *Systema naturae* of 1735. We might not think of it as a language, but Condorcet called it the first analytical scientific language for botany.[30]

Not everyone was enthusiastic about new analytical languages in science, however. The problem with Linnaeus's system was the value of the signs. Did Linnaeus's nomenclature accurately represent real forms of plants, or was it merely an artificial directory of names with no connection to the real essences of things? The most famous attack on Linnaeus's system was leveled by the Comte de Buffon in his 1749 *Initial Discourse on the Manner of Studying and Treating Natural History*. Much has been written about this famous "Introduction." It is a puzzle, because in it Buffon criticizes mathematics along with Linnaeus, and it is not apparent what mathematics has to do with taxonomic systems in botany. Moreover Buffon had been a great supporter of Newton's calculus and had translated Newton's *Treatise of Fluxions* into French with an introduction in 1740, therefore one would presume that Buffon was one of those lovers of analysis that Lagrange talked about and not a likely critic of the differential calculus.

In the "Initial Discourse" Buffon states explicitly that he regards Linnaeus's system as a language, but an unsatisfactory one. He accuses Linnaeus of "judging the whole by a single instance, to reduce nature to the status of petty systems which are foreign to her. . . . The final disadvantage of such methods" writes Buffon, "is that, in multiplying names and systems, they make the language of science more difficult than the science itself."[31] And further on he writes:

> For who does not see that whatever proceeds in such a manner cannot be considered a science? It is at the very most only a convention, an arbitrary language, a means of mutual understanding. But no real cognizance of things can result from it. . . . These various systems . . . have multiplied to the point that botany itself is actually easier to apprehend than the nomenclature which is merely its language?[32]

Buffon criticizes Linnaeus's system because it is artificial. The characters that Linnaeus used to identify plants overlook and obscure the real essential differences between plant forms. Linnaeus's analytical language misleads more than it informs.

In expanding his argument Buffon linked Linnaeus's artificial system with mathematics. In place of true philosophy, he claims, we now learn scientific skills which are artificial. "The skills which one would like to call scientific have taken [the place of philosophy]. The methods of calculus and geometry, those of botany and natural history— formulas, in a word, and dictionaries—occupy almost everyone. We think that we know more because we have increased the number of symbolic expression and learned phrases. We pay hardly any attention to the fact that all these skills are only the scaffolding of science, and not science itself."[33]

Buffon goes on to say that the truths in mathematics are only truths of definition. Mathematics tells us nothing new about the world, because it can only give back in another form what we have already put into it. Mathematical truths have the advantage of always being precise and conclusive, but they are also abstract, intellectual and arbitrary. It is true, says Buffon, that Newton's mathematization of the world system was a prodigious accomplishment, but the rest of the physical world will not lend itself to that kind of treatment.

The only other physical science that can be mathematized to the same degree is optics. As for rational mechanics, he argues that it is entirely mathematical, and not really part of physics. He writes: "The true goal of experimental physics is . . . to experiment with all things which we are not able to measure by mathematics, all the effects of which we do not yet know the causes, and all the properties whose circumstances we do not know. That alone can lead us to new discoveries, whereas the demonstration of mathematical effects will never show us anything except what we already know."[34]

Buffon is obviously unhappy with the method of analysis as a general method of scientific investigation. It is perhaps not surprising, then, that he and the Abbé Condillac were confirmed enemies and that the anti-mathematical and anti-analytical current that blossomed later in the works of Diderot, Rousseau, Bernardin de Saint-Pierre and Goethe often harks back to this criticism by Buffon.

Bernardin de Saint-Pierre is a good example of the anti-analytical sentiment popular in some quarters at the end of the eighteenth century. His pastoral novel *Paul et Virginie* was the most popular book to be published in France in 1788. It went through hundreds of editions in all European languages and it proved that the romantic sensibilities first aroused by Jean-Jacques Rousseau continued undiminished. Bernardin de Saint-Pierre had scientific pretensions. He had attended the Ecole des Ponts et Chaussees and had worked as a military engineer. His first literary success had been his *Etudes de la nature* (1784) which described the harmonies of Nature as the operation of Divine Providence. In 1792 Louis XVI appointed him Intendant of the Jardin des Plantes, declaring him to be a worthy successor to Buffon, and in 1795 he was admitted to the Institut de France. But Bernardin de Saint-Pierre was not fond of analysis and as a result was not looked upon with favor by his scientific colleagues. Perhaps Bonaparte gave the best explanation. When Saint-Pierre complained that his colleagues paid little attention to his works, Bonaparte is said to have replied: "Do you understand differential calculus, Monsieur Bernardin? No? Well then go and learn it, and then you will be able to answer your own question."[35] Napoleon, who appreciated mathematicians, was not willing to listen to Sainte-Pierre's criticism of analysis.

Critics of the method of analysis increased in number during the second half of the eighteenth century even as analysis went from triumph to triumph in physics and mathematics. The analysts dissected Nature, while Saint-Pierre and his idol, Rousseau, wanted to take Nature as one harmonious whole.

While some, like Buffon and Bernardin de Sainte-Pierre attacked the languages of analysis, others created them. The most obvious example of a new analytical language was the *Method of Chemical Nomenclature* introduced by Lavoisier and his supporters, the chemists Guyton de Morveau, Fourcroy, and Berthollet. This book appeared in 1787, right at the time when there was the biggest debate about method. It established the nomenclature that we follow in chemistry today and it replaced the old alchemical names such as "liver of sulfur," "butter of antimony," "Fuming liquor of Libavius," or Newton's favorites, the "Green Lyon," and the "stellate regulus of Mercury." The old names gave no indication of the composition of a compound. Lavoisier's nomenclature is, in itself, a kind of analysis, because it names chemical substances by their constituent parts.

Lavoisier's justification for the new chemical language followed Condillac's arguments very closely. He wrote in the introduction: "A well-composed language, adapted to the natural and successive order of ideas will bring in its train a necessary and immediate revolution in the method of teaching. . . . The logic of the sciences is . . . essentially dependent on their language."[36] By 1787 Condillac had been dead for seven years, but Lavoisier was obviously attempting to put Condillac's program into practice. Lavoisier's new nomenclature fitted only his system of chemistry. By accepting his nomenclature one had, of necessity, to accept his oxygen theory of combustion. After he had finished his experiments on heat with Laplace and had obtained the evidence that he felt he needed in order to destroy the phlogiston theory, he fixed his own oxygen theory with a symbolic and analytical language that left no room for phlogiston.

A third, and much less well known language, was created by Alexander von Humboldt, who described the geology of South America in an "Essay on Geological Pasigraphy" published in 1803. The word "pasigraphy" means a universal language and the late eighteenth century was replete with pasigraphies of all kinds created to improve the analytical process of reasoning and to reveal the common roots of all languages.

Humboldt's pasigraphy, however, was what we now call a vertical geological section. It appeared for the first time in his 1803 essay. As Humboldt described it, the purpose of his pasigraphy was "to present to the eye of the geognost the nature of the dominant rocks, . . . their direction and inclination . . . their thickness, the greatest and least heights at which they have been found, the absolute elevation of the mountains and valleys [in which they lie]."[37] Humboldt's pasigraphy was a new analytical language for geology.

The three scientific languages that I have mentioned, Linnaeus's binomial classification in botany, Lavoisier's system of chemical nomenclature, and Humboldt's geological pasigraphy, were all analytical, but none were mathematical. It is perhaps surprising that we find new scientific languages more commonly in natural history than in physics and that the impulse to quantify at the end of the century was particularly strong in natural history. Humboldt's pasigraphy was made possible by improvements in the science of barometric hypsometry, that is, the science of measuring mountain heights with the barometer. It was a science that had been created by naturalists, who explored the mountains to study their flora and geological structure. The first published list of mountain heights appeared in 1783. Without a base level from which to survey mountains, the only way to compare heights was with the barometer, and it was not possible to do this very precisely until the 1780s. This was especially true in the great expanses of the new world where it was impossible to determine mean sea level far from the coasts. The barometer was a fussy instrument, however, and in order to get accurate measurements it was necessary to make a careful analysis of the effects of temperature, the condition of the mercury (it was necessary to boil the mercury to expel absorbed gases that could spoil the measurements), the variations of **g** with the latitude, and other factors. Only careful measurement and testing in a great variety of circumstances resolved these problems. Laplace, a mathematician, finally put all of these variables together in a single barometric formula, but the work of determining and measuring the variables was the work of naturalists like Jean-André Deluc, William Roy, and George Evelyn-Shuckburgh.

Similar examples can be found in meteorology. Horace Benedict de Saussure, who made the first scientific study of the geology of the Alps, built the earliest hygrometer and measured the change in the vapor pressure of water with temperature. He also designed an early electroscope that was the basis for Volta's straw electrometer. (De Saussure also attempted to measure atmospheric electricity.) Volta also made elaborate quantitative experiments on the expansion of air with temperature. This kind of experimental work was highly quantitative, but it was not the preserve of mathematicians.

Susan Cannon has called this kind of activity "Humboldtian science" after its most distinguished practitioner, and has pointed out that it could not have existed before about 1770, because the available instruments were too crude. [38] It is not immediately obvious that what Cannon calls Humboldtian science is the same as Condillac's "analysis," but Humboldt's concern for the language of science suggests that it had some of the same elements.

Another aspect of language was closely related to quantification in physics. Condillac's claim that "we think only with the help of words" meant that not only physical objects, but also physical concepts had to be named. But in order to have good words it was necessary to have good concepts to which to assign the words, and those concepts were not available until about the same time that accurate measuring instruments became available, that is, until about 1780.

As mentioned above, in 1782 Laplace, Lavoisier, and Volta all worked together in Paris on atmospheric electricity while Lavoisier and Laplace were preparing to begin experiments with their new ice calorimeter. Lavoisier and Laplace built their calorimeter in 1782 because they had only then learned about specific and latent heat. Joseph Black had discovered (or perhaps we should say had created the concepts of) latent and specific heats in the 1760s, but he had not published them. Lavoisier and Laplace knew nothing about his experiments, and did not have a clear idea of specific heat until 1781. Carl Wilcke published his independent discovery of latent and specific heats also in 1781. Without the concepts of latent and specific heats, the ice calorimeter could not even be imagined. In this case the instrument had to wait for the concepts before it could be designed. Once Lavoisier and Laplace understood specific and latent heats, then they could begin quantitative calorimetry experiments. An analytical language of heat was not really possible until there were proper concepts to name.

It was just at this time, at the end of 1781, that Volta arrived in Paris and began working with Laplace and Lavoisier on electrification by evaporation. These experiments were made possible by Volta's extremely sensitive condensing electroscope. When he published a description of his electroscope in 1782, he wrote down for the first time the rule that the charge on a conductor is directly proportional to the capacitance and to the electrical "tension."[39] The science of electricity was far advanced by 1781, but quantitative measure had to wait until there was something to measure, and there was nothing to measure until natural philosophers had analyzed experience and had created concepts such as charge, capacitance, tension, specific heat, and latent heat, and had identified them with appropriate words. In this sense quantitative experiments, physical concepts and the language of science were all interdependent.

Of course one did not have to read Condillac in order to come up with the idea of specific heat. Joseph Black almost certainly got along quite well without reading Condillac. Yet I think the changes that we have observed in experimental and mathematical physics in the 1780s and the simultaneous attempt by philosophers to clarify the method of analysis were not accidental. They were seeking a common goal in their efforts to make analysis the single acceptable method in science. It was not a new method. Newton had understood it, and Condillac thought that in his study of analysis he was only following Newton's arguments to their logical conclusion, but it was new in what it produced. In particular it produced new instruments that made it possible to extend quantitative measure beyond astronomy and optics and into the study of heat, electricity, chemistry, and meteorology. It meant that the analytical methods of mathematics and mechanics could now be applied to a much wider range of experimental results. As we have seen, the method of analysis was not universally popular in the eighteenth century, nor was it employed exclusively by those with mathematical training. In France it was closely associated with the reform of language for good (in the opinion of

Lavoisier) or for ill (in the opinion of Buffon). Moreover the method of analysis never was satisfactorily defined, either in mathematics or in experimental physics. Nevertheless its ambiguity did not seem to lessen its importance in the eyes of those who employed it and it remained at the heart of the new quantitative physics.

NOTES

1. J.-L. Lagrange, *Méchanique analytique*, 4th ed. (Paris: Albert Blanchard, 1965), 1:i–ii.

2. Henry Guerlac, "Chemistry as a branch of physics: Laplace's collaboration with Lavoisier," *Historical Studies in the Physics Sciences*, 7 (1976):193–276.

3. J.-L. Lagrange, *Oeuvres de Lagrange*, J.-A. Serret, editor, 14 vols (Paris: Gauthier-Villars, 1882), 14:123–24.

4. Lavoisier and Laplace did not in fact obtain the degree of precision that they anticipated. Because the ice in their calorimeter held back the meltwater by surface tension and capillary action, their measurements of specific heat were approximately 10% too low. As it turned out, the method of mixtures gave better results than the ice calorimeter. See T. H. Lodwig and W. A. Smeaton, "The Ice Calorimeter of Lavoisier and Laplace and Some of its Critics," *Annals of Science*, 31 (1)(Jan. 1974):1–18.

5. For a description of these experiments and for citations to Volta's own comments see Guerlac, "Chemistry as a Branch of Physics," pp. 234–40, and John Heilbron, *Electricity in the 17th and 18th Centuries: A Study of Early Modern Physics* (Berkeley: University of California Press, 1979), p. 423.

6. While error calculation had become a concern of Lavoisier and Laplace, they did not systematically employ it in practice. In their joint "Mémoire sur la chaleur," which they read to the Paris Academy of Sciences in June 1783, Lavoisier gave his results in decimal fractions. This was an innovation, but Lavoisier showed no concern for significant figures. Guerlac, "Chemistry as a Branch of Physics," p. 253.

7. Heilbron, *Electricity in the 17th and 18th Centuries*, p. 480.

8. I. B. Cohen, *Franklin and Newton: An Inquiry into Speculative Newtonian Experimental Science* (Philadelphia: The American Philosophical Society, 1956), pp. 115–117.

9. Thomas S. Kuhn, *The Essential Tension: Selected Studies in Scientific Tradition and Change* (Chicago: University of Chicago Press, 1977), pp. 31–65.

10. Clifford Truesdell, "A Program Toward Rediscovering the Rational Mechanics of the Age of Reason," in *Essays in the History of Mechanics* (Berlin: Springer-Verlag, 1968), pp. 93–95.

11. A history of method which takes analysis as its central theme is David Oldroyd, *The Arch of Knowledge: An Introductory Study of the History of the Philosophy and Methodology of Science* (New York: Methuen, 1986). Chapter 1 treats the ancient tradition. More detailed studies of geometrical analysis are Jaako Hintikka and Unto Remes, *The Method of Analysis: Its Geometrical Origin and its General Significance*, Boston Studies in the Philosophy of Science XXV (Dordrecht: D. Reidel Publishing Company, 1974), Carl B. Boyer, "Analysis: notes on the evolution of a subject and a name," *Mathematics Teacher*, 47 (1954):450–62, and Paul Tannery, "Du sens des mots analysis et synthesis chez les Grecs et de leur algebre geometrique," in *Mémoires Scientifiques*, vol. 3 (Paris: 1915):158–69.

12. Arpad K. Szabo, "Working Backwards and Proving by Synthesis," in Hintikka and Remes, *The Method of Analysis*, pp. 118–26 and Francis M. Cornford, "Mathematics and Dialectic in the Republic VI–VII," *Mind* N.S. 41 (1932):37–52, 173–90.

13. Richard Robinson, "Analysis in Greek Geometry," in his *Essays in Greek Philosophy* (Oxford: Clarendon Press, 1969), pp. 1–15 and Harold Cherniss, "Plato as Mathematician," *Review of Metaphysics*, 4 (1951):414–425.

14. Hintikka and Remes, *Method of Analysis*, pp. xiii, 4, and 67.

15. Michael Mahoney, "Another Look at Greek Geometrical Analysis," *Archive for History of Exact Science*, 5 (1968):318–48. Without arguing for a later interpolation, Norman Gulley claims that Pappus described in the same passage two distinct methods, both of which he called analysis. ["Greek Geometrical Analysis," *Phronesis*, 33 (1958):1–14.]

16. Oldroyd, *Arch of Knowledge*, chapters 1 and 2, and Henry

Guerlac, "Newton and the Method of Analysis," in his *Essays and Papers in the History of Modern Science* (Baltimore: The Johns Hopkins Press, 1977), 193–215.

17. Isaac Newton, *Opticks*, 4th ed. (London: 1730), pp. 404–405.

18. Isaac Barrow, *Mathematical Lectures Read in the Publick Schools at the University of Cambridge*, John Kirkby, trans. (London, 1734), p. 28.

19. Guerlac, "Newton and the Method of Analysis," p. 195.

20. Guerlac, "Newton and the Method of Analysis," p. 6.

21. Oldroyd, *Arch of Knowledge*, pp. 66–67 and his article "Robert Hooke's Methodology of Science as Exemplified in his *Discourse of Earthquakes*," *British Journal for the History of Science*, 6 (1972):109–30. Also Mary B. Hesse, "Hooke's Philosophical Algebra," *Isis*, 57 (1966):67–83.

22. Voltaire, *Oeuvres complètes*, Louis Moland, ed., 52 vols. (Paris: Garnier freres, 1877–85), 22:423, 428.

23. Jean d'Alembert, article "Analyse," *Encyclopédie, ou Dictionnaire raisonné des sciences, des arts et des métiers*, (Paris: Briasson, David l'aîné, Le Breton, Durand, 1751–1780), 1:403.

24. Abbé Etienne Bonnot de Condillac, *Oeuvres philosophiques*, Georges Le Roy, ed., 3 vols. (Paris: Presses universitaires de France, 1947–51), 2:407.

25. Condillac, *Oeuvres*, 2:371.

26. Condillac, *Oeuvres*, 2:376.

27. Condillac, *Oeuvres*, 2:469.

28. Condillac, *Oeuvres*, 2.

29. In the remainder of this passage Kant explains the distinction: "The analytical method, so far as it is opposed to the synthetical, is very different from one that consists of analytical propositions; it signifies only that we start from what is sought, as if it were given, and ascend to the only conditions under which it is possible. In this method we often use nothing but synthetical propositions, as in mathematical analysis, and it were better to term it the regressive method, in contradistinction to the synthetic or progressive. A principal part of logic too is distinguished by the name of analytic, which here signifies the logic of truth to contrast to dialectic, without considering whether

the cognitions belonging to it are analytical or synthetical." *Prolegomena to Any Future Metaphysics Which Will be Able to Come forth as a Science*, 276n, from Lewis White Beck, *Kant Selections* (New York: Macmillan, 1988), p. 169. For more on this subject see Jaako Hintikka, "Kant and the Tradition of Analysis" in his *Logic, Language Games and Information: Kantian Themes in the Philosophy of Logic* (Oxford: Clarendon Press, 1973), chapter IX.

30. Marie-Jean-Antoine-Nicolas Caritat, marquis de Condorcet, *Oeuvres de Condorcet*, A. Condorcet-O'Connor and F. Arago, ed., 12 vols. (Paris, 1847–49), 2:342.

31. George-Louis Leclerc, comte de Buffon, "Initial Discourse" to the *Histoire naturelle*, in *From Natural History to the History of Nature: Readings from Buffon and His Critics*, John Lyon and Phillip R. Sloan, ed. and trans., (Notre Dame, 1981), p. 100.

32. Buffon, "Initial Discourse," p. 104.

33. Buffon, "Initial Discourse," p. 122.

34. Buffon, "Initial Discourse," p. 126.

35. Joseph Francois Michaud, *Biographie universelle*, 52 vols. (Paris: Michaud Frères, 1811–1828) vol. 40, p. 64. Perhaps Sainte-Beuve gave the best explanation for the popularity of Saint-Pierre's *Etudes de la nature*. He concluded that: "L'esprit était très-evéille aux idées nouvelles de science en 1784; la chimie, la physique, allaient changer de face par les travaux des Laplace et des Lavoisier. Si elles avaient paru dix ans plus tard, en 95 or 96, les Etudes eussent trouvé la nouvelle science déjà constatée et régnante, l'analyse victorieuse de l'hypothèse; en 84 elles purent obtenir, même par leur côté le plus faux, un succès de surprise et les honneurs d'une vive controverse." (C.-A. Sainte-Beuve, *Portraits littéraires* [Paris: Garnier Frères, (1862)], 2:125.)

36. Quoted from Thomas L. Hankins, *Science and the Enlightenment* (Cambridge: Cambridge University Press, 1985), p. 109. Two years later in the introduction to his *Traité élémentaire de chimie* Lavoisier wrote: "Nous ne pensons qu'avec le secours des mots; les langues sont de veritables méthodes analytiques; l'algèbre la plus simple, la plus exacte et la mieux adaptée à son object de toutes les manieres de s'enfoncer, est à la fois une

langue et une méthode analytique; enfin . . . l'art de raisonner se reduit à une langue bien faite." (Antoine Laurent Lavoisier, *Traité élémentaire de chimie* [Paris: Gauthiers-Villars, 1937], xxv.)

37. Quoted from Theodore S. Feldman, "Applied mathematics and the quantification of experimental physics: The example of barometric hypsometry," *Historical Studies in the Physical Sciences*, 15, pt. 2, (1985), p. 191.

38. Susan Faye Cannon, *Science in Culture: The Early Victorian Period* (New York: Science History Publications, 1978), p. 96.

39. Heilbron, *Electricity*, p. 457.

5

Teleology and the Relationship Between Biology and the Physical Sciences in the Nineteenth and Twentieth Centuries

John Beatty

> It is singular how differently one and the same book will impress different minds. That which struck the present writer most forcibly in his first perusal of the 'Origin of Species' was the conviction that Teleology, as commonly understood, had received its deathblow at Mr. Darwin's hands.
>
> T.H. Huxley

In his 1869 address, "The Aim and Progress of Physical Science," Helmholtz recounted a series of discoveries from Galileo to Newton, and finally to Darwin, all of which contributed to the "ultimate aim of physical science," which was "to reduce all phenomena to mechanics."[1] Darwin's role in this series of developments, as Helmholtz imagined it, had consisted in perhaps the ultimate extension of mechanistic thinking. For Darwin had succeeded in providing a mechanical alternative to teleological reasoning about the living world.[2]

The most striking thing about the living world, Helmholtz admitted, was the adaptation of organisms to their environments. Teleologists had urged that this sort of appropriateness could not be explained solely in terms of mechanical causes that are blind to their consequences. Organisms must have the characteristics they do at least in part because of the adaptive utility of those characteristics. But, according to Helmholtz, Darwinism made clear precisely "how adaptation in the structure of organisms can result from the blind rule of a law of nature."³

Helmholtz's mechanistic aspirations, and Darwin's, if we adopt the fairly common understanding of Darwin's achievement exemplified by Helmholtz, were a good deal more radical than those of many of the scientists that we associate with the rise of the mechanical worldview. Mechanists of the likes of Newton and Boyle never dreamed that nature could be understood wholly in mechanical terms. Complete understanding of nature would require knowing also the purposes that God had in mind when he employed this rather than that particular mechanism in the construction of the world. As Newton reasoned in the "General Scholium" that concluded the *Principia*, "mere mechanical causes" could not account for such things as the fact that the stars are so distant from one another. Rather, one would also have to take into account God's ends in designing the universe in this way. The good thing about having the stars so far apart, and especially so distant from our own solar system, is that they do not thereby collapse upon one another through their mutual gravitational forces. Rather more infamous is Newton's insistence that a mere mechanical explanation could not account for the origin of a solar system so orderly as our own. Only in terms of God's aesthetic ends could one account for the neatly nested concentric, and nearly uniplanar orbits of the planets and their moons, and for the common direction of their axial rotations and revolutions around the sun.⁴

But the inorganic world did not really present anything like the opportunity for teleological reasoning offered by the organic. Even Boyle, a staunch defender of the place of teleology within the mechanical worldview, had concerns about the usefulness of teleological reasoning in the physical sciences. In his "Disquisition about the Final Causes of Natural Things," he admitted that, "for my part, I am apt to think, there is more of admirable contrivance in a man's muscles than in (what we know of) the coelestial orbs; and that the eye of a fly is (at least as far as appears to us) a more curious piece of workmanship than the body of the sun."[5] And while he admitted that a purely mechanical account of the formation of different stones and metals might suffice, he would not allow the same for plants and animals.[6]

Kant, who was in so many other respects concerned to make sense of Newton, departed from him in this regard: not by denying the need for teleological reasoning in science, but by restricting it to the biological sciences. Kant distinguished between the nonliving realm of nature that we can hope to understand completely in terms of "mechanical principles," and the living world that we cannot; our understanding of the living world requires in addition the notions of purpose and design. It is absurd, he maintained in the *Critique of Judgement*, to expect that "another Newton will arise in the future who will make even the production of a blade of grass understandable by us according to natural laws which no design has ordered. . . . we must absolutely deny this insight to men."[7]

To be sure, Kant acknowledged, the formation of organisms and their parts is to be understood to some extent in terms of mechanical processes—"accretion" and the like— but the fact that the parts of organisms should take shapes so appropriate for their possessors, and that they should be placed so appropriately, "must always be estimated teleologically."[8]

Kant, in his style, considered purpose and design not so much a part of nature as of our cognitive apparatus; the notions of purpose and design were, for him, *a priori* "regulative ideas" that "guide [our] reflection upon nature" and make possible a unified conception of the otherwise very diverse living world.[9] Kant believed that we could hardly reflect upon purpose and design without invoking an intelligent Being whose purposes and designs are at issue. However, he also considered this additional inference to be "extravagant" as far as science itself is concerned and unnecessary as far the possibility of achieving a unified conception of the living world.[10] But what is most important here is just Kant's attitude toward the need for teleological, not strictly mechanical, thinking in biology, if not in the physical sciences.

Kant's attitude is nicely exemplified in his *Theory of the Heavens*, where he proposed a purely mechanical alternative to Newton's teleological account of the origin of the solar system (although he did not explicitly single out Newton for criticism on this occasion). However, Kant felt the need to preface the presentation of his nebular hypothesis with the acknowledgement that he foresaw no similar mechanical explanation of "even a single herb or caterpillar."[11]

The nebular hypothesis of Laplace received considerably more attention than Kant's, in part because of the (now well known) moral that Laplace drew from his achievement, namely, that "If we trace the history of the progress of the human mind, and of its errors, we shall observe final causes perpetually receding, according as the boundaries of our knowledge are extended" (Laplace did explicitly criticize Newton in this regard).[12] And thus it was to be expected (according to Laplace) that the origins of the various forms of life would also be explained one day in purely mechanical terms.[13] But it remained for such an account to be provided—much less generally accepted. And in the meantime, the need for teleological reasoning in the biological, if not the physical, sciences could still be defended.[14] For instance, that peculiarly Anglicized Kantian, William Whewell, while one of the nineteenth century's staunchest proponents of the search for God's ends, nonetheless discouraged such pursuits in astronomy and physics, the subjects of his Bridgewater Treatise. Citing Boyle as precedent, Whewell warned that the search for God's ends could "pervert the strict course of physical enquiry."[15] Such reasoning did play an ineliminable role in the sciences of life, however. In his *Philosophy of the Inductive Sciences*, Whewell argued that,

> This idea of a Final Cause is applicable as a fundamental and regulative idea to our speculations concerning organized creatures only. That there is a purpose in many other

parts of creation, we find abundant reason to believe, from the arrangements and laws which prevail around us. But this persuasion is not to be allowed to regulate and direct our reasonings with regard to inorganic matter, of which conception the relation of means and end forms no essential part. In mere Physics, Final Causes, as Bacon has observed, are not to be admitted as a principle of reasoning. But in the organical sciences, the assumption of design and purpose in every part of every whole, that is, the pervading idea of Final Cause, is the basis of sound reasoning and the source of true doctrine.[16]

So from the point of view of some of the most influential early mechanical philosophers, and many of their intellectual descendants, teleological reasoning had a place in science, though much more so, and according to some exclusively so, in the biological sciences. The pursuit of purposes actually served to distinguish quite neatly the practice of biology from that of astronomy, physics, and chemistry.

To the extent that Darwin succeeded in replacing teleological with mechanistic explanations in biology, then, he did away with one of the last (legitimate) remnants of teleological reasoning in the sciences, and also did away with a traditional way of distinguishing the biological from the physical sciences. In the process, he (would have) effected a radical change in, not just an extension of, the mechanical worldview of Boyle, and especially Newton.

But the question whether Darwin actually succeeded in replacing the teleological approach is a very difficult one, and I do not just mean in retrospect and from a purely philosophical point of view. My main concern here is with evaluations of Darwin by his contemporaries. After briefly introducing a few of the many varieties of teleological reasoning in nineteenth-century biology, and the nature of Darwin's alternative, I will then take up some of the very disparate perceptions that Darwin's supporters and critics had of his attitude toward teleology. Darwin was praised for abandoning teleology and criticized for pretending that it could be abandoned, but he was also criticized for being a teleologist and even praised for accommodating that approach.

In a recent paper that is similar to mine in spirit if not in substance, John Cornell raises the question of whether Darwin can reasonably be interpreted as the "Newton of the Grassblade" (referring to the passage from Kant discussed above). On the basis of a careful analysis of Kant's teleology, contrasted with more religious versions, Cornell suggests that Darwin succeeded (and self-consciously so) in undermining the more religious versions of teleology, but not necessarily Kant's. Summarizing his findings, Cornell notes, "Darwin's famed triumph over teleology must remain difficult to measure."[17] Exactly!

In the final section of this paper, I will jump ahead almost a century to evolutionary biology in the 1950s and 1960s in order to assess the Darwinian legacy in this regard. As we shall see, the question of whether Darwin was or was not a teleologist has not been resolved in the long interim. But leading Darwinians are nonetheless certain of the autonomy of biology, and on Darwinian grounds. Whether or not Darwin was a teleologist, he provided a way of drawing a distinction similar to that between pure mechanism and teleology, a distinction that, like the latter distinction, was used, and has continued to be used up to the present day, to distinguish the biological from the physical sciences.

DARWIN AND TELEOLOGY IN THE NINETEENTH CENTURY

In nineteenth-century biology (more generally, but at least in this context), teleology and mechanism were rarely characterized independently of one another, and rarely independently of the notion of chance as absence of purpose, or coincidence, or accident. As the German biologist Ernst Haeckel distinguished between the teleological and purely mechanical approaches,

One group of philosophers affirms, in accordance with its teleological conception, that the whole cosmos is an orderly system, in which every phenomenon has its aim and purpose; there is no such thing as chance. The other group, holding a mechanical theory, expresses itself thus: The development of the universe is a monistic process, in which we discover no aim or purpose whatever; what we call design in the organic world is a special result of biological agencies; neither in the evolution of the heavenly bodies nor in that of the crust of our earth do we find any trace of a controlling purpose—all is the result of chance.[18]

Examples of nineteenth-century teleologists so characterized are numerous. Consider for instance the German embryologist Karl Ernst von Baer. Von Baer could not fathom how the wonderfully adaptive characteristics of developing organisms could be the result of chance—mere coincidences, accidents. As an example, he pointed out how chicks still in the egg develop two hard spikes on the tips of their beaks at just the time when their backbones are well enough developed for them to move and stretch their necks.[19] By these movements, and with these two spikes, they break their eggs and emerge. Is the coincident timing of these phenomena just an *accident*? Surely the breaking of the egg and the emergence of the chick is the *purpose* of the appearance of the spikes. Sarcastically, von Baer formulated the alternative, mechanistic perspective of the phenomenon: instead of saying that the chick developed the spikes for the purpose of breaking the egg, the mechanist would say, "Because the hard spikes are there, it is possible for the egg to be broken from within."[20] But then why, von Baer pressed, do the spikes fall off just after the chick emerges, when they serve no further purpose?

There were many versions of teleological reasoning in nineteenth-century biology, of which I will discuss just a few. One involved the invocation of purposeful vital forces—von Baer was a proponent of this approach. He invoked vitalistic, organizational "types" to explain what physics and chemistry alone could not explain about the development and functioning of organisms, namely, the appropriate integration of the various physical and chemical processes that occurred.[21] Type-guided development was not strictly mechanical: the principles of physics and chemistry were controlled, if not violated, in the process. The materials and principles of physics and chemistry were, in this sense, means to an end—that end being the manifestation of the type. There had to be some way of guiding physical and chemical processes to an appropriately adaptive result. For, just as a productive laboratory needs a chemist, and not just chemicals and chemical principles, so too adaptively organized organisms must have some way of integrating their own various chemical processes.[22]

Among von Baer's detractors were mechanistic anti-teleologists, like Schwann, who was well known for his strictly mechanical account of cell development, which he conceived as analogous to crystal growth.[23] Other teleologists opposed the invocation of vital directing agents to explain the purposefulness of the organic world, but they were no less vigorous in their opposition to the notion that nature is the product of only coincidentally adaptive, strictly mechanical causes. Among them were the Kantian-inspired German biologists—recently labeled "teleomechanists" by Timothy Lenoir—who pursued mechanistic explanations of organic phenomena within a teleological framework.[24] They presumed that the mechanisms of life served purposes related to the generation and maintenance of life, and this assumption guided their mechanistic pursuits. For instance, Carl Bergmann and Rudolf Leuckart were guided in their investigation of the mechanisms of animal locomotion by the teleological principle that locomotion should occur in an efficient way.[25] But these teleomechanists were considerably more agnostic with regard to the nature of the causal efficacy of purposes than was their countryman von Baer. That is, while they considered the mechanisms of life to be subject to efficiency principles and the like, the nature of those constraints—for instance, whether they required vital forces, or whether they were instituted at the time of creation by God—was not considered by them to be susceptible to scientific investigation.

The version of teleology that predominated in the land of Darwin was the sort that identified purposes in the organic world with the purposes God had in mind when he fashioned living beings. The Reverend William Paley defended this approach forcefully at the very beginning of the nineteenth century in his immensely popular book, *Natural Theology.*[26] His line of reasoning is still familiar. On what grounds do we infer that a telescope, or any other obvious object of contrivance was indeed so contrived? Surely we infer a maker for such an object on the ground that its parts are so constructed and arranged as to suit their purposes relative to the overall purpose of the object. If a telescope can be so understood, then why not an eye? Like a telescope, the eye has a lens for the refraction of light rays to an area where the image is registered. And like a telescope, whose lens positions can be manipulated to focus on objects near and far, the lens of the eye also changes shape to focus on objects at different distances. Better even than a telescope, the eye has an automatic protective cover, and an automatic cleaning system, as well as a mechanism for directing it to objects of interest. The eye is clearly an object of contrivance, its purposefulness clearly a reflection of the purposes of its Designer.[27]

To understand nature in terms of no more than the interplay of mechanical forces, without acknowledging the purposes of its Architect, would be to construe the wonderful adaptations of nature as mere coincidences—the results of chance alone. But, as Paley reasoned,

> What does chance ever do for us? In the human body, for instance, chance, i.e., the operation of causes without design, may produce a wen, a wart, a mole, a pimple, but never an eye. Amongst inanimate substances, a clod, a pebble, a liquid drip, might be; but never was a watch, a telescope, an organized body of any kind, answering a valuable purpose by a complicated mechanism, the effect

of chance. In no assignable instance hath such a thing existed without intention somewhere.[28]

The nine Bridgewater Treatises published in 1839 represent the epitome of this particular theological, teleological, anti-chance view of nature.

Von Baer blamed this sort of theological teleology for much of the "teleophobia" of his time.[29] The theological brand of teleology had proven not only misleading, but positively silly, as was illustrated by the tale of the schoolmaster who explained the location of the world's major rivers by arguing that God had put them there to supply water to the world's major cities.[30]

Among those more positively influenced by theological teleology was the young Charles Darwin. Having pursued that line of reasoning for some time, he was then considered by many to have demolished it, the greatest blow coming with the publication in 1859 of his *On the Origin of Species*.[31]

Darwin himself at times conceived of his enterprise as an extension of mechanistic thinking analogous to the developments culminating in Newton's work. He quite self-consciously searched for a biological analogue to the law of gravitation. As he reflected early on in his working notebooks,

> Astronomers might formerly have said that God ordered each planet to move in its particular destiny. In same manner God orders each animal created with certain form in certain country, but how much more simple and sublime power let attraction act according to certain law, such are inevitable consequences—let animal be created, then by the fixed laws of generation, such will be their successors. Let the powers of transportal be such, and so will be the forms of one country to another—Let geological changes go at such a rate, so will be the number and distribution of the species!![32]

Jonathan Hodge and Michael Ruse have explained how Darwin also incorporated Newtonian methodological considerations—the sort supposedly exemplified in Newton's own mechanistic reasoning—into the defense of his theory of evolution by natural selection.[33] But let us go to the substance of the theory.

On Darwin's account of the unity and diversity of life, chance in the sense of absence of purpose, or chance in the sense of coincidence or accident, played a large role. Darwin usually invoked chance or accident specifically with reference to how new variations arise from time to time among the members of a species. Suppose, for instance, as Darwin did, that among the wolves of a certain area some speedier wolves happened "by chance" to be born—suppose, that is, that the original occurrence of these speedier wolves was independent of the fact that speed is beneficial for these wolves. Such speedier wolves would likely outreproduce and outsurvive the others. Assuming not only that they left more offspring but also that their offspring inherited their greater speediness, then a greater proportion of the next generation of wolves in this area would be speedier. And the proportion would continue to increase over time. Note that speediness would not increase among wolves if their circumstances did not make speed important to survival and reproduction, as, Darwin suggested, in the case of a certain form of North American wolf whose chief prey consists of sheep. Thus, wolves might diverge evolutionarily to the extent that their circumstances differ.[34] Extrapolating, Darwin proposed to explain how all forms of organisms come to have characteristics suited to their particular circumstances.

The important point for now concerns the particular role of chance in all of this. To say that advantageous variations arise by chance is to say that their occurrence is not occasioned by the fact that they promote the survival and reproduction of their possessors, though, of course, the persistence and increase in frequency of those characteristics does depend on their effects in their possessors. As Darwin more eloquently explained the role of chance variation,

[Evolution by natural selection] absolutely depends on what we in our ignorance call spontaneous or accidental variability. Let an architect be compelled to build an ed-

ifice with uncut stones, fallen from a precipice. The shape
of each fragment may be called accidental. Yet the shape
of each has been determined . . . by events and circumstances, all of which depend on natural laws; but there is
no relation between these laws and the purpose for which
each fragment is used by the builder. In the same manner
the variations of each creature are determined by fixed
and immutable laws; but these bear no relation to the
living structure which is slowly built up through the power
of selection.[35]

So, in other words, it is a matter of chance that there should
have fallen from the precipice a stone suitable for a particular use in a building. But of course it is not merely a matter
of chance that we find the stone put to that use. Similarly,
again, it is a matter of chance that a variation suitable to
the survival or reproductive needs of members of a particular species should arise within that species, but it is not
merely a matter of chance that variation should continue to
increase in frequency within the species.

This seemingly (misleadingly) simple concept nevertheless lent itself to widely different interpretations by Darwin's
contemporaries, especially with regard to the question of
whether it accommodated, or was a substitute for, teleological reasoning. As I mentioned earlier, Darwin was both
praised and criticized for abandoning teleology (see fig. 5.1),
but he was also praised and criticized for not doing so. I will
consider, briefly, examples of each of the four possible positions; suffice it to say that there are more examples of each

	For Promoting Teleology	For Undermining Teleology
Praised by	Gray	Huxley Helmholtz duBois-Reymond
Criticized by	Kölliker	von Baer

FIGURE 5.1. Darwin Praised and Criticized

position. Also, I hope it will be clear, in the context of this paper, why I am considering relatively more examples of one particular position than of the other three.

Among those convinced that Darwinism accommodated teleology was the American botanist Asa Gray. A longtime confidant of Darwin, Gray took issue with reviewers of the *Origin* who claimed that Darwin had repudiated teleological thinking; instead, Gray argued, "Darwin's particular hypothesis, if we understand it, would leave the doctrines of final causes, utility, and special design, just where they were before."[36]

> As to all this, nothing is easier than to bring out in the conclusion what you introduce in the premises. If you import atheism into your conception of variation and natural selection, you can readily exhibit it in the result. If you do not put it in, perhaps there need be none come out.[37]

This much Darwin would have been happy to acknowledge. What he objected to was the nature of the role that Gray had assigned to God in relation to evolution by natural selection, namely, the role of directing the occurrence of beneficial variations upon which selection could act.

> [W]e should advise Mr. Darwin to assume, in the philosophy of his hypothesis, that variation has been let along certain beneficial lines. Streams flowing over a sloping plane by gravitation (here the concept of natural selection) may have worn their actual channels as they flowed; yet their particular courses may have been assigned; and where we see them forming definite and useful lines of irrigation, after a manner unaccountable on the laws of gravitation and dynamics, we should believe that the distribution was designed.[38]

But, as Darwin rejoined, to reason thus was to miss the whole point about variation being a matter of chance. And this was essential to his notion of evolution by natural selection. Natural selection is rather "superfluous" if God just directs variation along beneficial lines.[39]

Darwin did not have to point out the essentially accidental aspects of evolution by natural selection to those like von Baer, who criticized him for having completely ignored teleology and abandoning the organic world to chance. For all its emphasis on chance variation, Darwin's account of evolution was patently absurd to von Baer.

That evolution by the natural selection of chance variations would ever lead to well adapted forms of life seemed to von Baer no more likely than that the residents of the isle of Laputa would ever succeed in developing knowledge in the manner reported by Gulliver. At the Academy of Lagado, in Laputa, a very mechanical approach to the acquisition of knowledge was supposedly being tried. The members of the Academy had inscribed words in all their grammatical forms on the sides of wooden dice. They had then connected the dice in such a way that the dice could be spun independently, and such that strings of words could be read off. After each spin of the dice, the strings of words were reviewed and those that formed sensible phrases were recorded. The dice were then spun again, with the hope that additional phrases would appear that could be conjoined with the former, and thus knowledge would be obtained. "The elimination of those that did not go together was . . . completely mechanical, and was completed much more rapidly than occurs in the 'struggle for existence.' "[40] Unfortunately, no one had taken Gulliver's account seriously, or they might have been in a better mind-set to appreciate Darwin's theory.

> For a long time the author of these reports was taken to be joking, because it is self-evident that nothing useful and significant could ever result from chance events. On the contrary, order must emerge as a complete whole at the outset, even though there might well be room for considerable improvement. Now we must acknowledge this philosopher as a deep thinker since he foresaw the present triumphs of science![41]

Just as it was incomprehensible to von Baer that knowledge could be generated in any other way than by choosing one's words carefully, so too it was incomprehensible to him that adaptation could be generated by chance.

Of course, the Laputans did not rely entirely on chance, and neither does evolution by natural selection. It was not just a matter of chance that sensible truths (vs. senseless phrases) would fill the pages of the books of the Academy of Lagado. The Laputans employed a selection mechanism in addition to their spins of the dice. Nor was it just a matter of chance that evolution by natural selection would produce adapted life forms. Unlike his countryman von Baer, the anatomist Albert von Kölliker saw considerable evidence of teleological thought in Darwin's work, and he brought Darwin to task for that reason:

> Darwin is, in the fullest sense of the word, a Teleologist. He says quite distinctly (First Edition, pp. 199, 200) that every particular in the structure of an animal has been created for its benefit, and he regards the whole series of animal forms only from this point of view.
>
> The teleological general conception adopted by Darwin is a mistaken one.
>
> Varieties arise irrespective of the notion of purpose, or of utility, according to general laws of Nature, and may be either useful, or hurtful, or indifferent.
>
> The assumption that an organism exists only on account of some definite end in view, and represents something more than the incorporation of a general idea, or law, implies a one-sided conception of the universe. Assuredly, every organism has, and every organism fulfills, its end, but its purpose is not the condition of its existence.[42]

In order to appreciate the nature of Kölliker's objections to Darwinism as outlined above, it is necessary to expand somewhat upon the alternative explanatory approaches available to biologists in the early to mid-nineteenth century. One important alternative, which we have yet to discuss, was the so-called "unity-of-plan" approach.[43] From this perspective, the characteristics of members of a species were best accounted for not so much in terms of the adaptive significance of those characteristics, but rather, first and foremost, in terms of resemblances between members of that and other species. Explaining the presence of a particular bone among members of a particular species—say, a species of bird—involved demonstrating first that this bone and the bone of another species of bird, perhaps also the bone of a fish, and perhaps also the bone of a mammal, are all, despite some differences, instances of the "same" bone, all "homologues." Thus, a common plan, repeated throughout nature, and underlying its variety, is revealed. As the French comparative anatomist Etienne Geoffroy St. Hilaire, one of the earliest systematic proponents of this view of nature claimed,

> It is known that nature works constantly with the same materials. She is ingenious to vary only the forms. As if, in fact, she were restricted to the [same] primitive ideas, one sees her tend always to cause the same elements to reappear in the same number, in the same circumstances, and with the same connections.[44]

If, as Geoffroy and others reasoned, organic form were so well tailored to the adaptive needs of organisms—if, for instance, fish were so well adapted to life under water and reptiles to life on land—then there would be no reason to expect such extensive similarities between them. As the English anatomist Richard Owen reasoned, there is no more reason to expect similarities in the organic world on the basis of Divine contrivance alone than appears among the products of human contrivance. Think how much more similar in terms of skeletal anatomy are birds, fishes, and land mammals than hot-air balloons, boats, and locomotives.[45]

Proponents of the unity-of-plan approach argued that the teleological approach placed too much emphasis on utility in nature. Unity-of-plan thinkers did not altogether deny that many characteristics of organisms are useful for the survival and reproduction of their possessors. But many characteristics are not. A favorite example was male nipples. The teleological approach also made little sense of the fact that organisms inhabiting very different environments (for example the temperate zones and the arctic), and hence supposedly adapted to very different environments, are often very much alike (as, for instance, in the case of the wolves of the temperate and arctic regions). Similarly, organisms in very similar environments can be very different. Thus, unity-of-plan thinkers advocated understanding the diversity of nature primarily as variations on a common theme, and only secondarily as adaptive variations.[46]

It may help to consider the approach in a little more detail, as it was propounded by Owen. In his books *Archetype and Homologies of the Vertebrate Skeleton* and *On the Nature of Limbs*, Owen developed the notion of what he called the "archetypical" vertebrate, an ideal creature that represented the essence of all vertebrates, without any of the specialized modifications characteristic of the various different real animals.[47] In addition to the archetypical vertebrate, there were also more specialized archetypical fish, birds, mammals, and humans. These, he claimed, were the patterns in accordance with which God originally intended that all creatures would be formed. These were the basic "blueprints," so to speak, for the formation of all living forms.

Owen himself did not believe that God had directly created each and every species as a modification of some archetype, though some unity-of-plan thinkers like the Swiss-American zoologist Louis Agassiz did.[48] Rather, Owen believed that God originally had the various archetypes in mind, and then endowed the world with mechanisms that had caused these themes to be repeated. God also endowed the world with mechanisms that adapted these repeated themes to their special environments, resulting in the formation of specialized characteristics serving specialized functions. Owen thus introduced a teleological sort of force into his unity-of-plan conception of the world. But he insisted that it was wrong to think that teleological reasoning was more basic in understanding organic form than the idea of conformity to a common plan.

It is with respect to the virtues of unity-of-plan reasoning that Kölliker perceived Darwin as just a teleologist, and faulted him for it. Darwin's emphasis on the respects in which natural selection resulted in useful combinations of traits made him too much of a utilitarian from Kölliker's more unity-of-plan point of view.

Kölliker, it is important to add, did not fault Darwin for the view that the multitude of species that have ever existed are ultimately descended from one or a few original species. What Kölliker objected to was the emphasis on natural selection in accounting for the modification of all the descendant species. He preferred to account for the various modifications in terms of a "law of development," by which original forms give rise, under particular conditions, to different versions of those forms, through a mechanism analogous to metamorphosis.[49] There may originally have been one original form, or several with different "developmental" capabilities: one that give rise to vertebrates, one that gave rise to invertebrates, etc. At any rate, it was the unity of plan exhibited in the living world, not the supposed utility of forms, that Kölliker thought most needed explaining. Actually, Darwin imagined himself to have found in evolution by natural selection the perfect compromise to the teleological and unity-of-plan approaches. In his scheme, common-ancestor species substituted for the archetypes or themes upon which a variety of different descendant species were generated. The variety of descendant species was generated by natural selection adapting each to the particular circumstances it encountered.[50] But this was not satisfactory to Kölliker, who complained that Darwin had not accounted for unity of type in terms of any law of nature guaranteeing that effect—as was the case for Owen's theory, and Kölliker's own—but rather "merely" as a matter of ancestry.[51]

The rebukes of Kölliker by "Darwin's Bulldog," T. H. Huxley, and by Darwin's schnauser, Haeckel, did not emphasize the distinction between the teleological and unity-of-plan approaches, nor Darwin's self-professed reconciliation.[52] I will concentrate on Huxley's response. Rather alarmed by Kölliker's appraisal of Darwin, Huxley retorted, "It is singular how differently one and the same book will impress different minds. That which struck the present writer most forcibly in his first perusal of the 'Origin of Species' was the conviction that Teleology, as commonly understood, had received its deathblow at Mr. Darwin's hands."[53]

To be a teleologist, as Huxley understood the enterprise, involved believing that organisms of different species take the different forms they do in response to a constructive agent that shapes them so that they *will* be suited, and *perfectly* suited, to the circumstances in which they find themselves. Darwinism, he argued, was quite at odds with this point of view. The suitability of a characteristic in a particular set of circumstances is relevant to the *persistence* and *increase* of that characteristic over the course of generations, but not to the *original occurrence* of that characteristic. An advantageous characteristic does not come into being on account of the adaptive advantages it offers in some set of circumstances. Moreover, the characteristics that do persist will be those that offer *better* survival and reproductive advantages to their possessors, not necessarily the *best possible*. In his own words,

> Cats catch mice, small birds and the like, very well. Teleology tells us that they do so because they were expressly constructed for so doing—that they are perfect mousing apparatuses. . . . Darwinism affirms, on the contrary, that there was no express construction concerned in the matter; but that among the multitudinous variations of the Feline stock, many of which died out from want of power to resist opposing influences, some, the cats, were better fitted to catch mice than others, whence they throve and persisted, in proportion to the advantage over their fellows thus offered to them.
>
> Far from imagining that cats exist *in order* to catch mice well, Darwinism supposes that cats exist because they catch mice well—mousing being not the end, but the condition of their existence.[54]

Huxley was a reductionist; his assessment of Darwin's role in purging biology of teleology fit in well with his view that biology was just a branch of physics and chemistry.[55] Similarly, the German reductionists Helmholtz and Emil du Bois-Reymond saw in Darwin's work the replacement of one of the last remnants of teleological thinking in science, and at the same time a significant extension of the boundaries of physical science. Helmholtz's interpretation of Darwin served as an introduction to my essay. I will add here only that it is indeed telling that an extended discussion of Darwin's accomplishments should have such a central place in Helmholtz's essay, "The Aim and Structure of Physical Science."

The notion that Darwin paved the way for the incorporation of biology into the physical sciences was also expressed forcefully by du Bois-Reymond, in his well-known essay, "Seven World Problems."[56] There were, as he explained there, seven questions yet to receive answers in terms of the physical sciences: the origin of matter and force, the origin of motion, the origin of life, the teleological character of life, conscious sensation, language and intelligent thought, and free will. As for the "apparently" teleological character of life, he remarked, "organic laws of formation cannot work adaptively unless matter was created with adaptive purpose in the beginning; and they are inconsistent with the mechanical view of nature." He continued, "The difficulty is, however, not absolutely transcendent, for Mr. Darwin has pointed out in his doctrine of natural selection a possible way of overcoming it, and of explaining the inner suitableness of organic creation to its purposes and its adaptation to inorganic conditions through a concatenation of circumstances operating by a kind of mechanism in connection with natural necessity."[57]

Helmholtz and du Bois-Reymond had considerably more faith in Darwinism than the majority of biologists of their time. As Peter Bowler and others have recently emphasized, Darwin's theory was anything but an overnight sensation, and was in fact just one of many rival theories of evolution in the late nineteenth and early twentieth centuries.[58] But du Bois-Reymond for one, whose reductionist interpretation of Darwinism made Darwin an ally, found the theory of evolution by natural selection appealing enough to cling to, and the image of clinging is not my own: "We might always, . . . while we hold to this theory, have the feeling of the otherwise helpless sinking man, who is cleaving to a plank that just bears him up even with the surface of the water. In the choice between the plank and destruction, the advantage is decidedly on the side of the plank."[59]

Did Darwin replace teleological thinking, or accommodate it? And in which direction did scientific progress lie? At least among Darwin's contemporaries, these were much disputed questions. So it was by no means immediately clear whether and/or how Darwin had affected the practice of biology vis-à-vis the physical sciences.

A CENTURY LATER

I am going to jump ahead now one whole century. It is the 1960s, and philosophically reflective biologists—for example, the evolutionary biologists Ernst Mayr and Francisco Ayala—are still at odds with regard to whether Darwin was or was not a teleologist.[60] But there is a deeper issue underlying this one, an issue reflected in the title of one of Ayala's papers on teleology, "Biology as an Autonomous Science," and with respect to this issue there is, in the sixties, considerable agreement, at least among evolutionary biologists.[61]

Consider the context. Molecular biology has been enjoying remarkable success, especially in the developments leading up to and surrounding Watson and Crick's discovery of the structure of DNA, and in the cascade of discoveries following from that. So significant are these developments that molecular biology is being called "the new biology," to the irritation of many biologists at the fringes of those developments, who are beginning to wonder whether the new biology is not actually just physics and chemistry.

A wave of antireductionist sentiment is growing in response to what evolutionary biologist George Gaylord Simpson calls the "DNA bandwagon effect."[62] The response is taking a variety of forms, one of the most influential of which is evolutionarily inspired, and vigorously promoted by evolutionary biologists of the stature of Simpson, Mayr, and Theodosius Dobzhansky. Central to the attempts of this Darwinian triumvirate to undercut the all-importance of molecular biology is their insistence that we not overlook the *evolutionary* perspective on life. My concern in this penultimate section is to convey briefly this evolutionary antireductionism, and thereby to show how it is the historical successor to the argument that the teleological character of life distinguishes the biological from the physical sciences. Darwin may have rid biology of teleology, but he did not, according to many of his staunchest twentieth-century defenders, thereby threaten the autonomy of biology.

Let us hear from Dobzhansky, Simpson, and Mayr in turn. As fundamental as the insights provided by physics and chemistry are, Dobzhansky argued, those fields do not constitute *the* fundamental perspectives on life. Underlying all characteristics of organisms and organismal processes are, to be sure, molecular structures and mechanisms. But their presence, in turn, must be understood evolutionarily, in terms of natural selection. For instance, as he explained, in order to understand how the process of respiration in vertebrates takes place, one needs to know something about the chemical properties of hemoglobin. But in order, in turn, to understand the presence of hemoglobin in vertebrates, one needs to understand the role it plays in maintaining vertebrate organisms, and hence the causes of its increased prevalence through natural selection within the lineages that led to today's vertebrates.

> It is often alleged that the molecular level of biological phenomena is the one in terms of which all other levels must be understood in "new" biology. This is why molecular biology is boastfully styled "fundamental." . . . It

is true that organismic level phenomena should be analyzed into molecular level components; it is equally true that the molecular components acquire meaning when viewed as the constitutents of organismic patterns and as products of the evolutionary development of the living world.[63]

Dobzhansky rarely missed the opportunity to stress the importance of evolutionary biology vis-à-vis molecular biology. His strongest single statement in this regard, the one so widely quoted—"Nothing in biology makes sense except in the light of evolution"—actually arose in the context of an article about how religious fundamentalism should not be allowed to interfere with the teaching of evolution.[64] How else, Dobzhansky asked, if not in terms of an evolutionary perspective, is one to make sense of all the rest of biology? How is one to make sense, in particular, of (then) recent findings concerning the universality of the genetic code and the method of its translation? Having summarized the major findings of several decades' work in molecular biology, Dobzhansky concluded, "I submit that all these remarkable findings make sense in the light of evolution; they are nonsense otherwise."[65] They are nonsense not only from a purely religious point of view but also from a purely molecular point of view. This is (or was certainly intended to be) a complete turnaround from the reductionist perspective—to think that evolutionary biology makes sense of molecular biology!

With equal enthusiasm, Simpson campaigned vigorously against the all-importance of molecular biology:

The rate of progress [in the various biological sciences] is uneven, and rapid advances take place now in one direction and now in quite another. Once a shove has been given in one direction, perhaps by a technological or conceptual breakthrough, perhaps by individual enthusiasm, perhaps by what seems pure chance, a band wagon effect ensues. Students flock to the accelerating front; money is poured into it; professional advancement, fame, and fortune follow it. That is only natural and is in one respect desirable, for the band wagon effect feeds into a circle by which the rate of discovery in some one field is indeed increased.

Fortunately, more than one aspect of biology may accelerate at the same time, and with the great increase in numbers of biologists not all crowding onto the same band wagon. The gaudiest band wagon just now is manned by reductionists, travels on biochemical and biophysical roads, and carries a banner with a strange device: DNA. There are, however, a good number of other, perhaps less gaudy band wagons, going down other, perhaps less rapid roads, under banners perhaps less vehemently saluted. At any rate, we salute them all, and yet may fear that these band wagons diminish travel on still other roads, which are falling into neglect but which are also essential to reach the destination toward which all are, or should be, traveling.[66]

Simpson acknowledged that physics and chemistry play an important role in explaining why organisms take the forms they do. Their role in that consists in elucidating *"how"* organisms are structured and how they function. But complete understanding of organic form requires also that we know *why* they are physically and chemically so characterized. For what reasons having to do with survival and reproduction do the organisms under investigation have these structures rather than some other? The why approach (together with Simpson's *"what for"* approach) involves, in short, considerations as to how natural selection has shaped species during their evolutionary histories.

The point about explanation in biology that I would particularly like to stress is this: to understand organisms one must explain their organization. It is elementary that one must know what is organized and how it is organized, but that does not explain the fact or the nature of the organization itself. Such explanation requires knowledge of how an organism came to be organized and what function the organization serves. Ultimate explanation in biology is therefore necessarily evolutionary.[67]

Although I have saved Mayr's antireductionism for last, it could well have been discussed first. The influence of his 1961 article, "Cause and Effect in Biology," on Dobzhansky's and Simpson's manifestoes is unmistakable. Interestingly, Mayr uses the concepts and terminology of molecular biology to mark the distinction that he uses to refute its all-importance. Biology, Mayr argues, is made up largely of two fields: "functional" and "evolutionary" biology. "The functional biologist deals with all aspects of the decoding of the programmed information contained in the DNA code of the fertilized zygote. The evolutionary biologist, on the other hand, is interested in the history of these codes of information and in the laws that control changes in these codes from generation to generation."[68] Mayr's point is simply this: a functional biologist ultimately traces back the structure and function of organisms of a particular species to the expression of their genetic material; the evolutionary biologist then takes up the issue of how, in the course of evolutionary history, organisms of the species in question came to possess the particular genetic material that they do. The latter issues often involve considerations of natural selection. Clearly neither field alone can provide a complete account of the organic world. They are entirely complementary.

CONCLUSION

One may resist identifying the distinctions pointed out by Dobzhansky, Simpson and Mayr with the distinction between mechanistic and teleological approaches—perhaps on the basis of one's conviction that Darwin did, after all, repudiate teleological thinking. Mayr, for one, has diligently denied that Darwinism is teleological, preferring the label "teleonomic."[69] Differences between the new and old distinctions aside, though, important similarities remain, even to the extent that they sometimes sound a lot alike. Consider, for instance, the additional terminology that Mayr uses to mark the distinction between functional and evolutionary biology, terminology he claims merely to have borrowed from another Darwinian.[70] He uses the term "proximate" causation to refer to the sorts of causes invoked in functional accounts, and the term "ultimate" causation to refer to the sorts of causes invoked in evolutionary accounts. Recall that it was common among early mechanical philosophers to distinguish mechanical from teleological explanations in terms of the similar-sounding distinction between "proximate" and "final" causes. The proximate/ultimate terminology enjoys wide use among contemporary biologists. (Of course, whether Mayr was aware of it or not, the term "ultimate" causation, referring to evolution, also plays a rhetorical role in promoting the importance of evolutionary biology relative to molecular biology.)

More important than just how similar they sound, though, the older mechanism (or proximate-cause) vs. teleology (or final- cause) distinction and the more recent how (or functional or proximate-cause) vs. why (or evolutionary or ultimate-cause) distinction have played very similar roles in distinguishing the biological from the physical sciences. Prior to Darwin, biologists did more than investigate the mechanical causes of organic phenomena, they sought teleological explanations as well. Physical scientists may have been obliged to restrict their enquiries to mechanical causes, but biologists were not. Whether or not the Darwinian revolution invalidated the search for teleological explanations, it did not restrict biologists to the same sorts of explanations as their physical science counterparts. Invocations of adaptive utility still play a major role in biology.

The Darwinian revolution changed biology, and with it, science. But it conserved a great deal as well, including central aspects of the mechanical worldview. It conserved the importance of adaptive utilitarian reasoning in the biological if not the physical sciences, and, along with that, it conserved the relationship between the biological and physical sciences.

The author is indebted to many historians of biology whose

names are included in the notes. Very special thanks go to Ernst Mayr and Shirley Roe, with whom I have greatly enjoyed discussing these issues over the years. To those familiar with the work of the late Dov Ospovat, it will be clear how much I have benefited from his contributions.

NOTES AND REFERENCES

1. Hermann von Helmholtz, "The Aim and Progress of Physical Science," [1869], in *Selected Writings of Hermann von Helmholtz*, Russell Kahl, ed. (Middletown, Connecticut: Wesleyan University Press, 1971), pp. 223–45, on p. 231.

2. Ibid., 237 ff.

3. Ibid., p. 238.

4. Isaac Newton, *Mathematical Principles of Natural Philosophy*, 3rd ed. [1726], trans. Andrew Motte [1729] Florian Cajori, ed. (Berkeley: University of California Press, 1934), vol. 2, pp. 543–544.

5. Robert Boyle, "Disquisitions about the Final Causes of Natural Things" [17XX], in *The Works of the Honorable Robert Boyle*, new ed. (London: Rivington, 1722), vol. 5, pp. 394–444, on p. 403. Discussed also in John Gillespie, "Natural History, Natural Theology, and Social Order: John Ray and the 'Newtonian Ideology,' " *Journal of the History of Biology* (1987): 1–49, on pp. 23–29. See also the more analytic treatment of Boyle's views on teleology in James G. Lennox "Robert Boyle's Defense of Teleological Inference in Experimental Science," *Isis* 74 (1983): 38–52.

6. Boyle, "Disquisition," pp. 403–404.

7. Immanuel Kant, *Critique of Judgement* [1790], trans. James C. Meredith (Oxford: Oxford University Press, 1928), "Second Part," section 14, p. 54 (section 75, p. 400 of Akademie ed.). For more general treatments of Kant's views on teleology, see J. D. McFarland, *Kant's Concept of Teleology* (Edinburgh: University of Edinburgh Press, 1970); Clark Zumbach, *The Transcendent Science: Kant's Conception of Biological Methodology* (The Hague: Nijhoff, 1984); Timothy Lenoir, *The Strategy of Life: Teleology and Mechanics in Nineteenth Century German Biology* (Dordrecht: Reidel, 1982).

8. See, e.g., Kant, *Critique of Judgement,* "Introduction," pp. 23–26 (pp. 184–186 of Akademie ed.).

9. Ibid., "Second Part," section 5, p. 26 (section 66, p. 377 of Akademie ed.).

10. Kant, *Critique of Judgement,* "Second Part," section 7, pp. 31–34 (section 68, pp. 381–384 of Akademie ed.).

11. Immanuel Kant, *Universal Natural History and Theory of the Heavens* [1755], trans. W. Hastie, in *Kant's Cosmogony* (New York: Johnson Reprint Corporation, 1970), p. 29.

12. Pierre Simon Laplace, *The System of the World,* [5th ed., 1824], trans. Henry H. Harte (Dublin: University of Dublin Press, 1830), p. 333; for the criticism of Newton, see p. 331.

13. Ibid., pp. 332–33. On the reception of Laplace's nebular hypothesis, and its relevance to the biological sciences, see Silvan S. Schweber, "August Comte and the Nebular Hypothesis"; Ronald L. Numbers, *Creation and Natural Law: Laplace's Nebular Hypothesis in American Thought* (Seattle: University of Washington Press, 1977).

14. See especially in this regard, Gillespie, "Natural History."

15. William Whewell, *Astronomy and General Physics Considered with Reference to Natural Theology* [1833] (London: Pickering, 1839), p. 355.

16. William Whewell, *The Philosophy of the Inductive Sciences,* 2d ed. (London: Parker, 1847), facsimile ed. (New York: Johnson Reprint, 1967), vol. 1, p. 628.

17. John F. Cornell, "Newton of the Grassblade? Darwin and the Problem of Organic Teleology," *Isis* 77 (1986): 405–21, on p. 421.

18. Ernst Haeckel, *The Riddle of the Universe* [1899], J. McCabe, trans., (New York: Harper and Brothers, 1900), pp. 273–74. Parts of this section of the paper come from my contribution to Gerd Gigerenzer, et al., *The Empire of Chance: How Probability Changed Science and Life* (Cambridge: Cambridge University Press, 1989), pp. 132–41.

19. Karl Ernst von Baer, *Reden gehalten in wissenschaftllichen Versammlungen und kleinere Aufsätze vermischten Inhalts* (St. Petersburg: Schmitzdorff, 1876), pp. 198–99.

20. Ibid., p. 199.

21. Karl Ernst von Baer, *Über Entwickelungsgeschichte der*

Theire: Beobachtung und Reflexion (Konigsberg: Bornträger, 1828–1837); von Baer, *Reden*. For more general discussions of von Baer's views on teleology and vitalism, see L. Blyakher, *History of Embryology in Russia: From the Middle of the Eighteenth to the Middle of the Nineteenth Century*, trans. anonymous (Washington, D.C.: Smithsonian Institution), pp. 339–64, 489–512; Lenoir, *The Strategy of Life*, pp. 72–95.

22. von Baer, *Reden*, p. 188.

23. Schwann, *Microscopical Researches*, pp. 186–215.

24. Lenoir, *The Strategy of Life*.

25. C. Bergmann and R. Leuckart, *Anatomisch-physiologische Uebersicht des Thierreichs* (Stuttgart: Muller, 1852); their most general discussion of the need for teleological reasoning is found on pp. 20–25. My discussion of Bergmann and Leuckart is based on Lenoir, *Strategy of Life*, pp. 172–94. See also William Coleman, "Bergmann's Rule: Animal Heat as a Biological Phenomenon," *Studies in the History of Biology* 3 (1979): 67–88.

26. William Paley, *Natural Theology, or Evidences of the Existence and Attributes of the Deity, Collected from the Appearances of Nature* (London: Fauldner, 1802).

27. Ibid., pp. 19–52.

28. Ibid., p. 46.

29. von Baer, *Reden*, p. 73.

30. Ibid., pp. 61–62.

31. See especially in this regard Dov Ospovat, *The Development of Darwin's Theory: Natural History, Natural Theology, and Natural Selection, 1838–1859* (Cambridge: Cambridge University Press, 1981); Neal C. Gillespie, *Charles Darwin and the Problem of Creation* (Chicago: University of Chicago Press, 1979); John C.Greene, *Darwin and the Modern World View* (Baton Rouge: Louisiana State University, 1961). John F. Cornell, "God's Magnificent Law: The Influence of Theology on Darwin's View of Natural Selection," *Journal of the History of Biology*.

32. Charles Darwin, "Darwin's Notebooks on the Transmutation of Species,"Gavin de Beer et al., eds., *Bulletin of the British Museum (Natural History)*, Historical Series 2 (1960–1961): 23–200, 3 (1967):129–76, on p. 53 of series 2 (pp. 101–102 by Darwin's page numbers).

33. Michael Ruse, "Darwin's Debt to Philosophy: An Ex-

amination of the Influence of the Philosophical Ideas of John F. W. Herschel and William Whewell on the Development of Charles Darwin's Theory of Evolution," *Studies in the History and Philosophy of Science* 6 (1975): 159–81; M. J. S. Hodge, "The Structure and Strategy of Darwin's 'Long Argument,'" *British Journal of the History of Science* 10 (1977):237–46.

34. Charles Darwin, *On the Origin of Species by Means of Natural Selection* (London: Murray, 1859), facsimile edited and with an introduction by Ernst Mayr (Cambridge: Harvard University Press, 1964), pp. 90–91.

35. Charles Darwin, *Variation of Animals and Plants Under Domestication* (London: Murray, 1968), vol. 2, p. 236.

36. Asa Gray, "Natural Selection not Inconsistent with Natural Theology," [1860], in *Darwiniana* [1876], edited with an introduction by A. Hunter Dupree (Cambridge: Harvard University Press, 1963), p. 119.

37. Ibid., p. 126.

38. Ibid., pp. 121–22.

39. Darwin, *Variation*, vol. 2, pp. 427–28. See also Michael Ghiselin, *The Triumph of the Darwinian Method* (Berkeley: University of California Press, 1969), pp. 131–59.

40. Karl Ernst von Baer, "The Controversy over Darwinism" [1873], David L. Hull, trans., in Hull, ed., *Darwin and his Critics* (Cambridge: Harvard University Press, 1973), pp. 416–25, on p. 419.

41. Ibid., p. 419.

42. Albert von Kölliker, "Ueber die Darwin'sche Schöpfungstheorie," *Zeitschrift für Wissenschaftliche Zoologie* 14 (1864): 174–86, on pp. 175, 178. These passages appear in the text above as they appeared in T. H. Huxley, "Criticisms on the Origin of Species" [1864], in *Darwiniana* [1893] (New York: Appleton, 1898), pp. 80–106, on p. 82.

43. With regard to differences between the teleological and unity-of-plan perspectives, see especially Toby Appel, *The Cuvier-Geoffroy Debate: French Biology in the Decades before Darwin.* (Oxford: Oxford University Press, 1987); Peter J. Bowler, "Darwinism and the Argument from Design: Suggestion for a Reevaluation," *Journal of the History of Biology* 10 (1977): 29–43; Ospovat, "Perfect Adaptation."

44. Etienne Geoffroy St. Hilaire, "Considérations sur les pièces de la tête osseuse des animaux vertébrés, et particulièrement sur celles du crâne des oiseaux," *Annales du Muséum d'Histoire Naturelle* 10 (1807): 342–65, on p. 344; quoted in Appel, *The Cuvier-Geoffroy Debate*, p. 89.

45. Richard Owen, *On the Nature of Limbs*, (London: Voorst, 1849), pp. 9–10; discussed in Ospovat, "Perfect Adaptation," p. 36.

46. See, for example, Louis Agassiz, *An Essay on Classification* (Boston: Little, Brown, 1857), pp. 9–17.

47. Owen, *On the Nature of Limbs*; Richard Owen, *On the Archetype and Homologies of the Vertebrate Skeleton* (London: Voorst, 1848).

48. Agassiz, *Essay on Classification*.

49. Kölliker, "Ueber die Darwin'sche Schöpfungstheorie," pp. 181–86.

50. Darwin, *Origin*.

51. Kölliker, "Ueber die Darwin'sche Schöpfungstheorie," p. 175.

52. Huxley, "Criticisms," Ernst Haeckel, *Generelle Morphologie der Organismen* (Berlin: Reimer, 1866), pp. 94–105.

53. Huxley, "Criticisms," p. 82.

54. Ibid., p. 85.

55. See also in this regard, John C. Greene, "The History of Ideas Revisited," *Revue de Synthèse* 4 (1986): 1–27.

56. Emil du Bois-Reymond, "The Seven World Problems," *The Popular Science Monthly* 5 (1882): 433–47.

57. Ibid., p. 438.

58. See especially Peter J. Bowler, *The Eclipse of Darwinism: Anti-Darwinian Evolution Theories in the Decades Around 1900* (Baltimore: Johns Hopkins University Press, 1983).

59. du Bois-Reymond, "Seven World Problems," p. 438.

60. Ernst Mayr, "Cause and Effect in Biology," *Science* 134 (1961): 1501–1506; Francisco J. Ayala, "Biology as an Autonomous Science," *American Scientist* 56 (1968): 7–221. Mayr was vs., Ayala pro. The issues in this section are dealt with in greater detail in John Beatty, "Evolutionary Anti-Reductionism: Historical Considerations," *Biology and Philosophy*, forthcoming.

61. See also Ernst Mayr's more recent "Is Biology an Autonomous Science?" in *Toward a New Philosophy of Biology: Observations of an Evolutionist* (Cambridge: Harvard University Press, 1988), pp. 8–23. Mayr's answer is, of course, "yes."

62. George Gaylord Simpson, *This View of Life* (New York: Harcourt Brace, 1964), p. 8. See also Ernst Mayr, "The New versus the Classical in Science," *Science* 141 (1963).

63. Theodosius Dobzhansky, "Are Naturalists Old-Fashioned?" *American Naturalist* 100 (1966): 541–50, on p. 545.

64. Theodosius Dobzhansky, "Nothing in Biology Makes Sense Except in the Light of Evolution," *American Biology Teacher* 35 (1973): 125–129.

65. Ibid., p. 128.

66. Simpson, *This View of Life*, pp. 113–114.

67. Ibid., p. 113.

68. Mayr, "Cause and Effect," p. 1502.

69. Ibid.; see also Ernst Mayr, "The Multiple Meanings of Teleology," in *Toward a New Philosophy of Biology*, pp. 38–66.

70. Ernst Mayr, *Growth of Biological Thought* (Cambridge: Harvard University Press, 1982), p. 68.

6

Newtonianism Before and After the Einsteinian Revolution

Joseph Agassi

PRELIMINARY REMARKS ON THE CONCEPT
OF APPROXIMATION

Approximations are tolerable, and even welcome, inaccuracies or imprecisions. The right degree of accuracy is what we agree is *a sufficiently good approximation*. There are both practical and theoretical limits to this accuracy. The degree of accuracy of a given measurement is decided by theory; but how is the accuracy or precision of the theory itself to be judged? Often the theory contributes much less to the inaccuracy or imprecision of a measurement than do its practical aspects. Sometimes there is a better theory to inform us as to the conditions under which a result guided by a lesser theory is a sufficiently good approximation. But what if there is not better theory? Which given theory is best? Is the best available theory good enough? Is the best theory available not also a mere approximation to a still better, as yet unknown, theory?

This is a difficult point. Given the truth, we can assess the degree of approximation to it attained by a theory or by measurement. On the assumption that a new theory is true, means are given for the assessment of the degree of approximation which the older one offers under some specific conditions. But if the truth is only asymptotically approached, then there is no way to decide. For what determines the degree of precision of a theory is not the idea of a better instrument or a better calculation, but the better theory which asserts *under what conditions* what result of the lesser theory is good enough an approximation. The claim that there is a true theory that is at the end of an asymptote is to say that were it known, then it would describe the conditions under which extant theories are sufficiently good approximations. Yet without it, we are left uninformed as to what these conditions are.

This difficulty is well-known among philosophers of science these days; it will not be aired here. Suffice it to notice that the very difficulty depends on the understanding of certain concepts, the concepts of approximate measurement, of a theory being sufficiently approximate to a later theory and of the approximation of the (as yet unknown and possibly unknowable) true theory. In brief, there may be no answer to the questions, "what is a good approximation?" and "what are the conditions under which one quantity or function is a good approximation to another?" But we are troubled by these questions, because we understand their implications. It is thus painfully clear why Einstein left open the question "is an ultimate theory possible?" but insisted on the imperative to hope for one. This also explains the great difficulty of developing his philosophy and the great incentive to stick to the older philosophy which suggests that Newtonian mechanics is the last word.

Joseph Agassi

A. NEWTONIAN MECHANICS BEFORE AND AFTER THE EINSTEINIAN REVOLUTION

Newtonian mechanics as it is understood today is not radically different from Newtonian mechanics as it was understood in the last days of its reign—before the Einsteinian revolution, especially in the formulation as presented by Poincaré.[1] Its ideas of the laws of motion and the laws of force are the same today as they were in the last century, and so are its demands that forces should be conservative and act at a distance in Euclidean space and absolute time. All of these ideas are maintained despite obvious reservations, whether regarding its claim that forces act at a distance or that space is Euclidean. This attitude needs examination, both as to the reasons behind it, and as to its real meaning. Here we encounter a disagreement: thinkers differ, but most of the examination of this attitude takes place outside physics, whether it is physics as it is taught or as it is practiced in research. Within physics there is remarkable unanimity. The old theory is accepted as a good approximation to the new theory in certain specified cases, and in these cases it is often easier to use the old theory than the new one. Consider, for example, the common practice of presenting gravity as if it were constant rather than variable. For a wide variety of studies of events that take place near the surface of the earth, it matters not whether we take it to be constant, as did Galileo, or a variable, as did Newton. The same holds for the invariability of time intervals, and of inertial mass, which Newton has taken as constant and Einstein did not.

The concern of the present discussion is with the *differences* between Newtonian mechanics today and Newtonian mechanics in its last days. The most important one is, possibly, the current accent on the invariance of Newtonian mechanics to Galilean transformations, an idea which is post-Einsteinian. It is hard for a physicist today, schooled in the characterization of a theory by the groups of transformations to which it is invariant, to realize that this approach

147

is hardly hinted at before the advent of general relativity, or at least before the Lorentz transformation replaced the Galilean transformation. Indeed, the clearest presentation of Newtonian mechanics as invariant to Galilean transformations is in Einstein's classic *The Meaning of Relativity*. But even that idea is not quite absent from the classical presentation of Newtonian mechanics; it is adumbrated in the principle of inertia, as we call Newton's first law. Now that principle is worded in two different ways, as if they were one. Put as the principle of uniform motion under no force, it is a special case of the dependence of acceleration on force; seen this way Newton's first law is a special case of Newton's second law. Put as the law concerning the indifference of a system's behavior to any uniform motion, it is fully Galilean invariance.

The absence of recognized measures of degrees of difference or of criteria of similarity or dissimilarity makes it hard to judge how much the current view of Newtonian mechanics differs from the view of it held a century ago. It does seem clear, however, that some differences are rather marginal. For example, though we are more used now to derive the description from a Lagrangean or a Hamiltonian function than was customary a century ago, this is of little significance. It seems to be quite generally true that there is little significance to the differences in wording of the theory today, from that customary a century ago.

This is not an obvious fact; it is not always the case that a theory remains intact through the ages. On the contrary, as Einstein has observed, it takes a long time before a theory is crystalized into a canonic version.[2] Thus, one struggles with Newton's own texts, and even with later ones, and the difficulty in reading the old texts is not merely a matter of different notation. There are differences in the intended meaning and in some significant perceptions of important characteristics of the theory. As to these perceptions, whereas we view Newton's theory of gravity as a theory of the conservative field *par excellence*, he himself thought otherwise. And since he considered his theory to be absolutely true, he assumed that now and then God Himself interferes in the solar system (in the system of the world, to use his locution) in order to supply the missing energy and ensure the stability of the solar system. As to intended meaning, the standard example is Newton's wording of his law of inertia, the first law. It has no special symbols and it is worded in reasonably modern English, yet those not steeped in Cartesian physics will doubtless find it either redundant (as merely a special case of the second law) or hard to comprehend. No such difficulties impede the reader of the late-nineteenth-century texts on Newtonian mechanics; after Helmholtz, Hamilton, and Jacobi, the picture was quite clear. Not that there was unanimity as to the different aspects of the theory, but the disagreement concerned the status of the theory, not its contents.

The main dispute over the status of Newtonian mechanics in the last half-century of its heyday was begun by Kelvin and Tait and continued by Ernst Mach.[3] It concerned the concept of force: can it be eliminated from the theoretical description? The situation was problematic and open to dispute, yet the disputed material itself was clearly enough understood.

That the situation remained controversial to the last, but that the controversy was clear enough, can be compared to the situation of the present dispute about quantum paradoxes and wave-particle duality. The difference is a complex matter, but at least we can say that with Hamilton's version a canonic formulation of Newtonian mechanics was at hand, whereas quantum mechanics received its canonic presentations only recently—even though John von Neumann offered one in the early thirties, soon after the different versions of the theory were shown to be more-or-less equivalent.[4] The dispute between Einstein and Bohr was about not only the status of quantum mechanics, but also what the theory actually says. This is much more disturbing. Nowadays, because of the agreement on a canonic presentation, that dispute seems to be a matter of history. There were similar debates in the seventeenth and eighteenth centuries over what

Newton's theory of gravity said; these have been aired in the classic historical studies of Alexandre Koyré.[5] In the nineteenth century, there was no such dispute about the theory.

Assuming that Newtonian mechanics itself has undergone no alterations since Hamilton and Jacobi, one cannot help notice that it looks these days remarkably different from the way it looked a century ago. It has not been changed, but its historical settings have. And the historical setting of a scientific theory has two aspects. One is informative content and thus concerns metaphysics or intellectual framework, as well as related scientific theories; the other concerns the status and methods of science in general. The informative content of the theory will be discussed in the next section, while the methodological aspect will be reserved to the section which follows it.

B. NEWTONIAN MECHANICS WITHIN AND WITHOUT NEWTONIANISM

The *locus classicus* of the definition of the Newtonian system was the opening passage of the classic essay of young Hermann von Helmholtz, "On the Conservation of Force," of 1847.[6] Ernst Cassirer, unaware of field theories (and, anyway, controversies within science were beyond his imagination), quoted this passage as the expression of the spirit of the age. In that famous passage Helmholtz said that no physical theory can be deemed satisfactory unless it assumes the existence of nothing other than particles interacting at a distance according to Newton's three laws. If a theory does not comply with this requirement, he adjudicated, it may be accepted as an empty, though possibly useful, mathematical formula, but not as a meaningful informative theory. This judgment of Helmholtz is not simply a statement about Newtonian mechanics. Rather, it is the claim that no theory should be taken as satisfactory unless it fits the Newtonian system, even though it may be used in a tentative way or even taken seriously.

The distinctness of the Newtonian system from Newtonian mechanics is discussed in detail in I. Bernard Cohen's classic *Franklin and Newton*, in which the usefulness of the Newtonian system or intellectual framework for Franklin's researches is illustrated in fascinating detail.[7] The *system*, of course, neither follows from nor entails any specific theory. To be within the Newtonian system, or within its framework, a scientific theory has to include Newton's laws of motion and, in addition, a law specifying the dependence of some force(s) on the distance between some sets of particles. But the Newtonian system, and any theory within it, are logically independent. In particular, from the obedience of Newtonian gravitational forces to Newton's three laws it does not follow that all forces do the same. (Thus the forces in electromagnetism are not central; they are conservative only globally, not locally.) Nor does it follow from the demand that gravitation should act as a conservative Newtonian force that it should vary as the inverse square of the distance between interacting bodies. It was well known before Einstein, and noted by Ernst Mach in his popular *On Mechanics*,[8] a work which helped lead Einstein to conclude that the Newtonian formula for gravity can be slightly modified by postulating that it propagates with the speed of light or by the addition to it of a small factor varying as the inverse cube of the distance, and that this modification has the advantage that it can explain the secular variation of Mercury's observed orbit from its calculated orbit. The added factor must be so small as to have little effect except on the planets nearest the Sun.

The demand of Helmholtz that all theories of physics be squeezed into the Newtonian mold shows both a liberal and an intolerant or unfriendly side. It was liberal of him to recognize such theories, to admit them to the club; it was intolerant of him to grant them only the status of visitors, until they came to "behave" in accord with the accepted rules.

The insistence by young Helmholtz on an illiberal rule, however, was not idiosyncratic. Being in no position to legislate, he was merely voicing, in the opening of his essay, an opinion shared by his readers. Why then did he reaffirm it? Presumably because he was going to violate the accepted code, and he was anxious to look conservative so as to get away with his heresy. He was anxious to allow field theories on this pretext, that since in the Newtonian system forces come in equal and opposite pairs so that their sum is always zero, the field theory of conservation of force can be admitted by the orthodox Newtonian. Though this solves one problem, concerning the conduct of the young Helmholtz, it raises another new and more difficult one, concerning most of the Newtonians in the last century: why were they so rigid in their reluctance to consider non-Newtonian theories as to need the reassurance of Helmholtz?

The intolerance exhibited by Helmholtz was much less stringent than that prevalent in his society. This explains the rejection of "On the Conservation of Force," which he then published privately. The official excuse, as recorded in the historical literature, is that it was suspected that he was a *Naturphilosoph*, a disciple of the philosophy of Schelling and Hegel, which (rightly or wrongly) was reputed to be highly anti-scientific.[9] This has to raise an eyebrow; what in that essay could possibly incur the suspicion of conformity to the metaphysics of these philosophers, and is this suspicion valid?

The answer is that historians of science are unwisely apologetic for science and for the great scientists of the past; they do not wish to belabor past scientific mistakes. In addition, many historians of science are guilty of praising what is *obviously* praiseworthy, whether through lack of courage, competence, or originality. The question that is raised here is not discussed in the literature: why was Helmholtz suspected of following the *Naturphilosophen*? My own opinion is as follows.

The accusation that one is a *Naturphilosoph* is but the accusation that one is a deviant. Now it is generally true that all deviants from a given orthodoxy share certain ideas in their deviation from the same orthodoxy, and furthermore the establishment likes to place all deviants in the same basket and to dismiss them all on the same grounds. This is particularly true of *Naturphilosophie*, since its exponents were notoriously ignorant, especially of mathematics. This can hardly be said of Helmholtz, but to claim that he was a deviant was more easily sustainable. What then was his deviation? As I have noted, it was that he attempted a compromise between the established Newtonian system of action at a distance and the newly evolved field system. He did so by finding something they shared. The law of action and reaction guarantees that the sum total of all active forces is zero. Hence within Newtonianism, forces conserve. The idea that forces conserve, the theory of the conservation of force, advocated by Helmholz as an orthodox Newtonian idea, was, however, something introduced by the field theorists. This, of course, is an excuse, and it is too superficial; the reason field theorists claimed that forces conserve is that the magnitude of a force is unchangeable, that the magnitude of each force is constant, that forces are capable of transformation of quality (for example, an electric force can become magnetic), but not of quantity. In Newtonianism, by contrast, forces cannot transform (since each force is the quality of a specific matter), but their magnitudes change with the change of distance between interacting bodies!

One then wonders why so much maneuvering was necessary. Why not offer the Newtonian and the field systems side by side and let the audience decide for itself? To understand this, we must move from the informative content of science to its *status*. Briefly, the dogmatic refusal to notice scientific mistakes, even important scientific mistakes, was rooted in a faith in the claim for certitude of empirical scientific theories. The faith in certitude imposed a rigid insistence that all scientific theories are absolutely true.

C. RATIONALITY AS SCIENTIFIC PROOF

We come now to the most important difference between the accepted picture of Newtonian mechanics then and now: in its time it was deemed to be absolutely true, while it is now deemed to be merely approximately true. This distinction invites some explanation before any elaboration on the matter of nineteenth-century Newtonianism can be undertaken.

At the outset, an observation of great importance from the field of the sociology of science is called for. Today the status of a "scientific theory" is usually given by physical scientists to any theory which is considered "verified." This This happens either when a theory is properly verified or if it is an approximation—presumably to another theory which is considered properly verified. More accurately, a theory is often called "verified" when the predictions based on it are sufficiently approximate to the information accrued in its empirical test. Now any valid verification of a theory is valid once and for all; a verified theory is true for all times. Hence there is in fact a significant difference between a verification and an approximation; many extant theories, such as Newtonian mechanics, are surprisingly good approximations; but they are not properly verified, once and for all. In all probability, no theory is properly verified, once and for all. Yet it is futile to demand of scientists that they not call an approximately true theory "verified." They know the distinction between a mere approximation and a verification (or "confirmation" or "empirical support"). Yet, they do not shrink from imprecise and inadequate language. This is similar to, but rather more serious than, their continued use of the word "atom" for particles that are by now known not to be atomic in the original Greek sense of indivisibility. Of course their interest is in physical theory, not in the theory of science.

So much for these observations from the sociology of science. One need not censure physicists for their showing only passing interest in this question of when a theory can be said to be empirical or "verified." The question is difficult, and it is a constant source of controversy; but it is central to the philosophy of science. It is advisable then to take it slowly, and to go to the history of science before tackling it. In order to understand the history of physics, perhaps one need not know the answer to our question. To understand the history of science, one can make do with the observation that among physicists the accepted answer to the question has undergone a serious change. This is very important, since it is impossible to comprehend the views physicists had a century ago on the status of Newtonian mechanics without seeing that they took it to be properly verified, not a mere approximation. This is, of course, an observation from the social history of science, and it would be a bit difficult to examine it empirically. Yet the evidence for it is very strong and was never contested.

This is not to say that historians of physics are in the habit of drawing attention to the fact that physicists have radically altered their view on the nature of science. Historians are not prone to stress the fact that in the last century most physicists declared a theory to be scientific only if it had been verified, yet today they are prepared to consider an approximation to a verified theory scientific as well. On the contrary, most historians of science have allowed their readers and audiences to conclude that throughout the ages their own views of the nature of science, and their own answer to our question, are always shared by most scientists, or at least by most able scientists since the onset of the scientific revolution. The important exception is Pierre Duhem, at once a physicist and a great philosopher and historian of science. Duhem emphatically declared that his view of science contradicted the received view. According to the received view, he observed, scientific theories are empirically verifiable; in his own view, there is no such thing as empirical verification. He went so far as to say that, quite possibly, we should one day deviate from Newtonian mechanics. This was heresy, and to his fellow physicists Duhem was suspect. He was eager to go to Paris, and finally he was invited to go there—but only as a historian of science. He took offense and rejected the invitation.[10]

Our concern here, of course, is with history, not physics. We might then have hoped to be enlightened by Duhem concerning the history of views about the status of science. Alas!

Duhem was a partisan for his view that scientific theories are mere intruments for prediction, because he wished to reinstate the Roman Catholic view of science as devoid of any metaphysical system; he wanted to reinstate Aristotle, in accord with Catholic tradition. Furthermore, since Duhem wrote before the Einsteinian revolution, and since he was opposed to Einstein to his dying day in 1918, he is of no help to us in the question of the impact of Einstein on the way we view Newtonianism.

Most studies of the history of Newtonian mechanics leave in their readers the very confused impression that there is no disagreement between the followers of Newton and those of Einstein. It is no accident that many historians of physics, including reputable physicists such as Sir Edmund Whittaker, declare Newtonian mechanics to be absolutely true in their work on the history of classical physics, but only approximately true in their work on the history of twentieth-century physics. This carelessness is itself of little importance, but it hinders the explanation of the rigidity researchers showed in the pre-Einstein years. This, however, the majority of historians of science do not notice at all, since their works present science and its history as perfect. But the history of humanity shows that nothing is perfect. It is interesting to note the rigidity of our predecessors, and for that we have to realize that they almost all accepted the theory that rationality requires proof and that the rationality of empirical science is that of empirical proof. The general view current at the time among scientists, philosophers, and the educated public was that rationality equals proof and that science is the paradigm of rationality. Even anti-scientific philosophers did not venture to contest this popular claim, and they gave vent to their anti-scientific or anti-rationalistic views by concocting odd and new forms of proof (or of what they called proof).

This is not true of twentieth-century thought, where one may find hostility to proof and also the view that we may be rational without proof. Prominent among the philosophers hostile to reason in general and to science in particular is Martin Heidegger. He said that scientific rationality is confined to science and that scientific truths concern technology exclusively so that science carries no real intellectual force. It is poetic truths, Heidigger added, that really signify.[11] We should remember that this was not so in the last century, when scientists and philosophers alike endorsed as a matter of course the view that thought is either rational or capricious, and that rational thought is characterized only by proof.

Unless we see that for physicists before the Einsteinian revolution of 1905, science is the body of properly verified theories, we shall not be able to appreciate the difference in research methods before and after that revolution, and we shall not be able to see its depth and the difficulty of its execution. The place to begin is with the difficulty that the revolution had to surmount, both as to theory and as to methodology. Thus we turn to the traditional attitudes toward approximations, given the view that scientific theories are verifiable.

D. APPROXIMATIONS THEN AND NOW

Once we declare a theory verified and hence true, there is no difficulty considering approximations to it. The idea of approximation to the truth is not new, of course, as it is highly intuitive (and as "old as the hills"). It fascinated Galileo, who asked if there can be a general measure of proximity to the truth.[12] Newton developed powerful methods of approximation that are still used in number theory (not in analysis or physics, where the Taylor expansion opened much more powerful techniques). In the present context, it is scarcely possible to ignore Newton's perturbation method, of which he was rightly proud. His theory explained (with the aid of his perturbation method, it should be added) the minute deviations of the outer planets from their Keplerian orbits. He considered this a powerful empirical argument in favor of his theory. In any case, as is well known, since the many-body problem of Newtonian gravity remains unsolved, there is no escape from the need to employ approximate solutions to it.

Nevertheless, approximately true scientific theories have been considered inferior at best. For example, Galileo's discussion of Archimedes' law indicates clearly that he saw it as absolutely true, whereas we know it to be only an approximation. The Aristotelean theory of gravity assigned to bodies both gravity and levity, in measures which depend on the heavy and the light elements that they contain. The simple refutation of this seems to be the floating of a stone or metal container, placed on the water in such a way that it does not fill with water. Aristotle paid notice to this fact in the very end of *On the Heavens* and declared that it is the shape of the container that is responsible for the floating, very much like the needle which floats on the water if placed carefully enough so as not to break the surface of the water. To put it in modern terms, the floating body may be subject to surface tension in addition to the forces of gravity and of levity.

So much for Aristotle's theory. It is a terribly bad one, even though some popular contemporary historians and philosophers of science, notably Thomas S. Kuhn and Paul K. Feyerabend, are now trying to rehabilitate Aristotle as a physicist, in the context of his own time. But if a body is subject to gravity and levity alone, it cannot be subject to gravity and levity and surface tension too. Aristotle repeatedly amended his theories by small corrections and modifications, forgetting that logically the correction, however small, renders the uncorrected theory false. This is an important point which should constantly stay in focus during the present discussion. Furthermore, there is little doubt that ascribing surface tension to a metal boat but not to a wooden boat is suspect. Aristotle simply did not work his theory out in detail; it is remote from the truth. Nevertheless, surface tension does exist, and the floating body will be subject not only to the force given by Archimedes' law, but to surface tension as well. Hence Archimedes' theory does not hold, and is refuted in cases where surface tension is important, as in the case of a floating needle. Finally, we cannot accept Archimedes' wording of his law, nor Galileo's: a floating body is no more weightless than a book resting on a table! But all this does not reduce Archimedes' law to the level of the theory of Aristotle. This, quite contrary to the view of Kuhn and Feyerabend, is why we teach Archimedes but not Aristotle in elementary physics classes.

Galileo was the first to recognize the following simple fact: that the admission of a correction to a theory, even if it is very small and even if it appears only under very rare circumstances, renders the uncorrected version of the theory false, though it may be a very good approximation to the corrected version. This is a great discovery, and it must increase our admiration for Galileo's rigor and tenacity and his intellectual courage.

Galileo placed much weight on the absolute correctness of Archimedes' law, and chiefly because Archimedes theory is presented axiomatically.[13] We may remember that Galileo became a Copernican because he was an Archimedean and recognized that although Aristotle's theory of gravity conflicts with Copernicanism, Archimedes' theory does not. He therefore attempted to prove that Archimedes' theory is absolutely true.

His proof is very simple: since Archimedes' theory is axiomatic, is must be true. Hence, when a needle with a specific gravity greater than that of water floats, it must expel a quantity of water greater than its volume. What Galileo assumed, in effect, is that a repulsive capillary force increases the volume of water expelled by the needle so as to enable it to float in accord with Archimedes' law. I hasten to add that neither surface tension nor capillarity were understood at the time in the same way they are today. In any case, Galileo declared his view not to be empirically refuted. Now clearly Archimedes' law is not absolutely true, as it does not, for example, fully describe the case of the floating needle. Historians of physics attempt to vindicate Galileo, thereby unwittingly granting recognition to some of his accusers. We should not view him as in need of defense.

When considering Galileo's theory of falling bodies, one need not be exceptionally clever to see that taking his theory of gravitational acceleration as constant implies a flat earth. Indeed, when he takes the limit of the inclined plane as a horizontal straight line he more than implies it—he makes it patently clear. There is, of course, no difficulty in viewing a small part of the surface of the earth as flat. The mathematical considerations are largely trivial, but interesting. The Galilean parabolic path of a projectile is an ellipse with one focus in infinity; taking the center of the earth as infinitely far away does make the earth flat. This kind of reasoning is now very familiar—recall that Einstein's radiation theory of 1917 was an exercise of deriving Bohr's equations from simple considerations with temperatures taken to the extreme. But it was not easy for Galileo's contemporaries to see this, and even Newton had to explain it in some detail in his *System of the World*.

What is surprising is that Galileo insisted that his theory is absolutely true! It was easy for Newton to see it as approximately true, as he taught that his own theory was absolutely true. It is much harder to view the last word as only an approximation to some unknown, perhaps unknowable truth.

E. PROBLEMS WITH APPROXIMATIONS

While physicists need not address philosophical problems, they cannot avoid philosophy altogether, even when their philosophical views are inadequate and superficial. One can hardly expect them to be indifferent to the theory of rationality and to go on doing their job, leaving all considerations of rationality to philosophers. The idea of the Age of Reason—that they should officiate as their own personal philosophers—is still very popular among physicists; and to some extent this is simply unavoidable: one has to have some idea of the meaning and worth of one's activities.

In the middle of our own century the opposite idea has evolved; the idea that philosophers should cease studying

Joseph Agassi

the aim and meaning of science and, taking the worth of science for granted, should leave the details to scientists. This view deprives scientists of all feel for the idea that science has value beyond its predictive value; predictive value suffices only for the power worshippers who admire science from afar. To the scientist, the predictive value of science is important because he wants to know that science is rational; it is not that prediction is valued more than rationality. In any case, scientists must at the very least have some feel for the idea that science is of value—and for more than the predictive efficacy that suffices for the power worshippers who admire science from afar. Scientists want to know that science is rational, and this is why they take predictive value as critical; it is not that they value it above rationality. As long as rationality is identified with verification, verification is what they crave. And the theory of rationality as verification either means empirical proof in the sense of the last century, and so is known to be erroneous, or it is vague.

This is not to demand from scientists a new theory of rationality, for it is unreasonable to expect a typical working scientist to develop such a theory, but it is to say that one is needed, and that a good theory of rationality may be of help to scientists also.

For even today there is no easy way to alter the theory of rationality of science. Our choice of options is very limited and none is comfortable. As a result of the Einsteinian revolution it is known that the classical theory of rationality as proof is quite inadequate. This requires that we develop a new theory to answer the following question: what theory or theories do physicists consider empirically verified, and why, and with what justice? The classical alternative to verification is the view that scientific theory is a mere instrument for prediction. Supposing this to be true, how do we construct a system of the world? We have the option of having no system of the world at all, of sticking to whatever commonsense view we happen to possess, of taking up one upon the faith of our forefathers, or of taking one capriciously or at random. None of these options is to be seri-

161

ously entertained, especially given our understanding of how powerful a means of fostering research such a system can be. When physicists praise a theory for its predictive value, they rarely suggest that there is no cognitive value to the theory beyond its instrumental value—the exceptions being the religious or the otherwise metaphysically committed physicists, whose committments bar them from taking physical theories literally.

What other options are there? Two are to be found in the literature: the so-called inductivist option, the approximationist option, and possibly their conjunction or disjunction.[14] Many philosophers hold the inductivist view. When they attempt to apply it to science proper, however, it becomes painfully clear that they also use the approximationist approach. It is clear, then, that the inductivist option alone is not seriously entertained. As to the combination of the inductivist and approximationist approaches, it raises two problems. First, is it necessary to have both? Second, is it possible to have both? The need to have both is felt because of the feeling that the latest theory justifies its predecessors, which are approximations to it, but that the latest theory needs an inductive justification. As to the question of compatibility, it is linked to the question of the status of an approximate theory. On this, Galileo's observation still stands, and waiving it is to fall back on Aristotelean confusions. An approximation, as he observed, however good it may be, is still false, since it contradicts the theory it approximates.

The theory of science as successive levels of explanation each inductively derived from their predecessors is now in total jeopardy. It is bad enough, as Karl Popper observes, that one admits the need to have rules not validated by logic. To have rules *contrary* to logic is too much. But things go further. Leave induction as a bad job and agree with a number of classical thinkers such as Galileo, Descartes, Kant, and William Whewell, that initially a scientific idea is but a figment of the imagination. What makes it more than that, what elevates it from the status of sheer fantasy to the status of science, is its explanatory power; science is a series of theories each of which explains its predecessors. This idea is very appealing and rarely questioned, but the popular theory of explanation of one law as mere deduction of it from another is invalidated by the theory of approximations. For if Kepler's theory contradicts Newton's, and Newton's contradicts Einstein's, then the theory of explanation forwarded by Whewell, Hempel, and Popper is false (Popper has modified his theory to account for approximations, but has not withdrawn his original theory). Where does that leave us? We can say that a theory is scientific when it is validated; when, in particular, it has passed a severe test.

This is the theory of Whewell and Popper. Whewell considered this sort of validation a complete verification. Popper attempted to present this as a version of approximationism. The new theory, he says, has to undergo a crucial test against the old and prove better than its predecessor both as an approximation and as an account of the facts. Yet the two are not identical; the claim that one theory is a better approximation to the truth, whatever it means, also means that any future crucial tests between the two will go the same way. That is to say, once we establish that the planetary orbit deviates from its Keplerian ellipse in accord with Newton, then we know that there will never be an observation of an orbit that will show a clear preference for Kepler over Newton. It is, of course, quite possible that in a contest between two theories not all crucial experiments will go the same way. In this case we shall, of course, look for another theory which will do better than both. But if experience has gone only in one direction, are we justified in expecting that it will continue to go that way?

In the last century, Helmholtz explained why we seek to verify experience.[15] If experience agrees with a given theory systematically, he said, then it may be accidental, or due to the theory being true. The more evidence we have, the more likely it is that it systematically verifies the theory because the theory is true. This lovely argument has been superseded by Einstein, who gave us a third option: it may be an accident and it may be due to the theory being true, as

Helmholtz observed, but it may also be due to the theory being approximately true. Yet once we say Newton's theory is successful because it is an approximation to Einstein's, we cannot finally avoid acknowledging that Einstein's is successful because it approximates yet another, still unknown theory. Does this series of theories, this infinite regression, converge? We have no reason to think so!

The difficulty concerns the complexity of the very concept of approximation. Two of its elements are intertwined and commonsensical; one of them, on the other hand, is very sophisticated. The two are of the more-or-less characteristic and of the under-most-common-circumstances characteristic. That is to say, an approximate result resembles the true result (a) when its quantitative measure is close enough and (b) when the conditions are of the common variety. That the two are intertwined is easy to see: under some conditions the inaccuracies can be magnified to any desired degree. All this is straightforward and known from the very early days of the scientific revolution.

The third aspect of the concept of approximation is irksome. Whereas the first two aspects relate the approximate to the true, the third correlates degrees of approximation and these allow series of approximations—perhaps even without ever achieving the truth. This is both sophisticated and problematic. The problems it relates to are now being aired in the philosophical literature, and the situation does not seem to be sufficiently under control.

As long as we operate with the concept of approximation *to the truth*, the difference between the true and the approximate is merely technical. As long as the true is in hand or even around the corner, the approximate is a shortcut for some technical problems or a shortcut to the achievement of the truth. This is how Galileo looked at things and this is how things remained until the Einsteinian revolution. Things look very different when we have two false alternative theories and we do not purport to have the true alternative to them. How do we judge which of the two is nearer to the truth? Is there a reason to assume that there is a complete ordering of these theories? The answer seems to be no.

What this amounts to in concrete cases is that we either claim that relativity is true or that we do not know if our contention that it is nearer some unknown truth is at all significant. To indicate the enormity of the problem, it remains to show that the complete ordering of theories is impossible.

This is an abstract consideration. There is yet a more abstract one, and it is that if we have no guarantee that the language we are using is capable of expressing the whole of God's truth about the universe, then it is meaningless to talk about the ultimate truth, the ultimate reality. But this invites total arbitrariness within science. To make the argument less abstract, let us consider some examples. In the last century, as we know, the theories of electricity that postulated action at a distance were much more popular among physicists than field theories—and for the reason that they seemed to be more in conformity with Newtonianism. The forces were not conservative, and that worried researchers, but fields of force worried them even more. There was a series of theories of electrodynamics: Coulomb's theory, Weber's, Ritz's theory, and more.[16] Except for Coulomb's theory, they have all ceased to be commonly recognized series of approximations to current theories. Why is this? If we should take Galileo's theory of gravity to be an approximation to Kepler's, we cannot take Kepler's as a possible approximation to Galileo's. In the cases of Newton and Einstein we could have viewed matters either way and needed experience to guide us. In his original 1905 paper introducing special relativity, Einstein showed that Newtonian kinetic energy is an approximation to the relativistic form.[17] But insofar as the truth of the theory is concerned, it could equally have been the other way around.

In other words, only when we take verifications to be indicators as to the possible increased proximity to the truth can we take the idea of approximate truth seriously. But this is when we have no justification but mere indications. And indications can easily mislead: they belong to heuristics, not to epistemology. They may help us direct our thoughts, but have no knowledge claims attached to them. This surely requires a revolutionary new view of rationality. The very best we have is still not so refined—debugged—that we may take the problem as solved. We can thus easily imagine how hard it was in the last century to relinquish the classical theory of rationality, at a time when it looked as if Newtonian mechanics would never be refuted by empirical evidence.

F. APPROXIMATIONISM AND RESEARCH

The pre-history of approximationism is rich and the discussion of it cannot be pursued here. One example should suffice: at times Newton declared the views of his predecessors, Kepler and Galileo, approximations to his own theory, and at other times he declared them absolutely true.[18] This is of course very embarrassing. Newton also declared that the Copernican system (the sun in the center of the universe) is absolutely true, even though it is not clear that it is consistent with his own system. This case is less embarrassing since it was declared to be true only in absolute space, and it is not clear what role absolute space has in his system (other than to account for rotations).

Without discussing these embarrassments, we may notice that what disturbed Newton was his realization that a theory which is an approximation to the truth, no matter how good, is false, and that no falsehood is provable, so that he was pressed by the theory of rationality as proof to declare Galileo's or Kepler's theory absolutely true, or not rational! Of course the way out is to devise a new theory of rationality, but to expect this even from Newton would be anachronistic.

It is a strange circumstance that the great defect of scholasticism is in its pretense that approximate theories are identical with the theories they approximate, that the great contribution of Galileo to the scientific ethos was his discovery of this contradiction, and that the great Newton confused the issue. Newton was not, of course, the only one.

Leibniz attempted to deduce Newton's theory of gravity from Kepler's laws, as did Newton's friends Colin McLaurin and Henry Pemberton, following Newton's lead here in accord with the way they understood the *Principia*. When the anti-scientific *Naturphilosoph* Hegel used this to prove that (the Englishman) Newton had plagiarized from (the German) Kepler, national pride and the rescue of the honor of science forced thinkers to have the issue clarified. It was by no means an easy job, however. It was performed by the great mid-nineteenth-century philosopher of science William Whewell, who attacked Hegel by proving that the two theories, Kepler's and Newton's, were logically equivalent only for a two-body system.[19] The proof is in the deduction of a contradiction from their equivalency in the three-body system. This is in keeping, we recall, with Galileo's idea that the modified version of a theory is inconsistent with the unmodified version of the same theory. This fact is very hard to accept, because of not only the Aristotelian tradition, which still has a deep hold on us, but also the identification of the rational with the correct. Many philosophers of science are still ignorant of this point, despite its elaboration with admirable clarity in the writings of Karl Popper.

The difficulty concerned the very idea that a theory may be scientific even if it is false and in contradiction to a scientific theory which is true. Whewell himself struggled with this idea, and his philosophy got entangled and was soon rejected as heresy. Clearly the alternative open to thinkers in such a situation is to opt for approximations as scientific regardless of the contradiction between the latest theory and its approximation; yet this is fraught with difficulties too.

Physicists of the last century did not articulate such difficulties; they simply did not concern themselves with approximations. Thus when Wilhelm Weber developed a modification of Ampere's theory of currents acting at a distance, he declared Ampere's theory plainly a prejudice![20] James Clerk Maxwell was more generous. When he rejected the advanced potentials of Lorentz, he proposed to offer Lorentz's theory, as a consolation prize, the status of an approximation. Yet Maxwell clearly thought too poorly of that prize. For when he attempted to calculate the field energy of Newtonian gravity he found out quickly that the energy is divergent. He was very disturbed and decided not to continue the study of gravity; it did not occur to him to study the possibility that the Newtonian gravitational force acts not quite at a distance! It is not that he could not imagine that possibility, for it was precisely the possibility proposed by Michael Faraday, whom he deeply admired. The possibility simply seemed to him too revolutionary to be entertained seriously.

Excluding the possibility that Newtonian mechanics is but an approximation to the truth, it might seem that there is no choice but to stay within the Newtonian system. This, however, need not be so; one may choose any system that one thinks is consistent with it. This explains Helmholtz, and Kelvin and Tait, and Maxwell as well. These were all followers of the field system who were hoping to show that both the field and Newtonian systems were coherent with the Cartesian system. This gave them some freedom, but not much, since it forced the advocates of field theories to reconcile this idea with the Newtonian system—which, as Kelvin and Maxwell explained in great detail, can only be done in the context of aether theories. This is why the problem of aether drift was deemed so central in the physics of the end of the Newtonian era.[21]

All this is not to say that Einstein developed approximationism in 1905. On the contrary, in "On the Electrodynamics of Moving Bodies" he came closest to holding the view that physics is better off without a system—a view which he repeatedly called, somewhat humorously no doubt, the sin of his youth. The fact that his 1905 paper contains a formula for approximation does not of course make it revolutionary. The revolutionary aspect of it is that it dethrones Newtonianism and places the field system firmly and independently on the map. Of course Einstein did not view the field system in 1905 in the same way he did toward the end of his life. Rather, in his paper on the photoelectric effect of the same year, he saw Maxwell's electromagnetic field theory as an approximation to some future photon theory of light. In each of the three celebrated papers of 1905 he spoke of approximation in the sense of approximation to the truth. Yet in each he had no theory which the current and established theories were supposed to approximate. This showed great daring.[22]

It transpired only decades later, mainly through the insights of my former teacher Karl Popper, whom Einstein greatly encouraged and with whom he often agreed, that Einstein's work was pregnant with revolutionary ideas in methodology and in metaphysics no less that in physics.[23] And, as with the physics, they have raised in these fields too, more problems than solutions.

NOTES AND REFERENCES

1. Henri Poincaré held a conventionalist philosophy of physics and presented dynamics in its field potential form, to render it as similar to electromagnetic theory as he could. The similarity is discussed in his philosophical writings (*Science and Hypothesis, Science and Method, The Value of Science* [collected as *The Foundations of Science*, Washington D.C.: University Press of America. 1982]), and illustrated in his astronomical and mechanical monographs. The similarity was, however, very partial and superficial, of course. His opposition to Lorentz's contraction stems from this effort to stress the similarity. See A. I. Miller, "Of Some Other Approaches to Electrodynamics in 1905," in *Some Strangeness in the Proportion: a centenary symposium to celebrate the achievements of Albert Einstein*, Harry Woolf, ed. (Reading, Mass.: Addison-Wesley, 1980), pp. 66–91, where Poincaré's conventionalism is presented briefly, on pages 68 and 87, in a one line text and a one line note (n. 6). Consequently, the paper tends to condemn rather that explain. Readers interested in Poincaré should not miss Jaki's *Uneasy Genius*, cited in note 10, below.

2. See Albert Einstein, *The Meaning of Relativity* [1921], 5th edition, Princeton: Princeton University Press, 1970. Gerald Holton notes that Einstein did not call his theory "the theory

of relativity"; he preferred the title of "invariantentheorie" ["theory of invariants"]. See Gerald Holton, "Einstein's Scientific Program: the Formative Years," in Harry Woolf, *Some Strangeness in the Proportion*, 49–63, p. 57. It is interesting to compare this essay with that of Abraham Pais, "Einstein on Particles, Fields, and the Quantum Theory," *ibid.*, pp. 197–265, especially Section 8. One should also compare pp. 226 and 472 on Einstein's anticipation of de Broglie's material wave equation.

3. See Max Jammer, *The Concept of Force: a Study in the Foundations of Dynamics* (Cambridge: Harvard University Press, 1957). See also my *Faraday as Natural Philosopher*, Chicago: University of Chicago Press, 1971, pp. 75, 114, 154, 194, 205, and 312.

4. On the alleged equivalence of Schrödinger's equation and matrix mechanics, see Mario Bunge, *Foundations of Physics* (New York: Springer-Verlag, 1967), pp. 249 and 253, and his *Philosophy of Physics* (Dordrecht: Reidel, 1973, p. 113). The latest on the formal aspects of the theory, Michael Redhead, *Incompleteness, Nonlocality, and Realism* (Oxford: Oxford University Press, 1987), presents the quantum formalism in its Dirac and its von Neumann formulation; the two earlier versions have proved too partial to serve as frameworks, and there is no proof of equivalence without some framework, of course. Schrödinger proved that the eigenvalues of his equation are the same as the eigenvalues of matrix mechanics in the (nonrelativistic) cases which were central in the time, but already the next step, Born's application of Schrödinger's equation to scattering phenomena, makes the equivalence very limited, as scattering is continuous and matrices are necessarily discrete.

5. Alexandre Koyré, *From the Closed World to the Infinite Universe*, (Baltimore: Johns Hopkins University Press, 1957).

6. Hermann von Helmholtz, *On the Conservation of Force, Scientific Memoires Selected from the Transactions of Foreign Academies of Science; Natural Philosophy*, Tyndall and Francis, eds., London, 1853. See also Ernst Cassirer, *The Problem of Knowledge: Philosophy, Science and History Since Hegel* (New Haven: Yale University Press, 1950), pp. 87–88.

7. I. Bernard Cohen, *Franklin and Newton: An Inquiry Into Speculative Newtonian Experimental Science and Franklin's Work*

in *Electricity as an Example Thereof*, (Philadelphia: American Philosophical Society. 1956).

8. See my *Faraday as a Natural Philosopher*, pp. 209 and 330, for the suggestion that no force acts at a distance, and Ernst Mach, *The Science of Mechanics: a critical and historical exposition of its principles* [1883] (Chicago: Open Court Publishing Co., 1902), pp. 334–35, for an endorsement of this view and a report of Paul Gerber's explanation (1898) of the secular motion of the perihelion of Mercury assuming that gravity propagates with the speed of light.

It is well-known that this result is incorporated in the general theory of relativity. Some philosophers of science have concluded from the fact that Einstein reported being confirmed by this result that when he worked on his theory he was not aware of the perihelion advance. See my *The Gentle Art of Philosophical Polemics* (La Salle Ill.: Open Court, 1988), p. 346. Philosophers of science are bound to be increasingly entangled in scholasticism until they come to see that confirmation in technology is regulated by law; see my *Technology; philosophical and social aspects* (Dordrecht: Reidel, 1985). In science, confirmation is either heuristic or an increase of explanatory power, or, as in a crucial experiment, a refutation in disguise. See my *Science in Flux*, (Dordrecht: Reidel, 1975).

Mach's readiness to criticize Newton's mechanics (see, for example, pp. 245, 492, 507, 535, 562–3, and 570ff) despite overwhelming confirmation and his readiness to see it modified, is what made Einstein suggest that he could discover the special theory of relativity. Einstein thought that the greatest obstacle to the development of that theory was the unwillingness to deviate from Newtonianism. This is, indeed, the point of the present essay. Yet Mach's attitude was not as fallibilist as Einstein's, despite some clearly fallibilist statements of his. See my *The Gentle Art of Philosophical Polemics*, pp. 21–32 on Mach.

9. Sir Edmund Whittaker, *A History of Theories of the Aether and Electricity, Vol. 1, The Classical Theories*, revised and enlarged edition (London: T. Nelson. 1951, p. 183).

10. See Stanley L. Jaki, *Uneasy Genius: The Life and Work of Pierre Duhem* (the Hague: Nijhoff, 1984), p. 374.

11. Any discussion of Martin Heidegger is bound to be con-

troversial. The question here is not whether Heidegger advocated a theory of poetic truth, but whether or not he promulgated the view that since poetic truth is higher than the prosaic, scientific-technological truth, the one is a legitimate constraint on the other. When one stays on the surface, one may evade this question, as does George Steiner, in his relatively lucid and umcomplimentary *Martin Heidegger* (New York: Viking Press, 1979), p. 146. But Laslo Versenyi cannot afford this luxury, as his book is dedicated to *Heidegger, Being, and Truth* (New Haven: Yale University Press, 1965, 1984), where being is the poetic mode and truth is either in the inferior prosaic mode or in the lofty poetic mode. He quotes the master (p. 41) to say that "we must presuppose truth" and declares that this cannot be taken literally [since it is a travesty on the truth]. Yet later (p. 92ff) he clear sides with the master in the claim that art is superior to science and that artistic judgment of truth overrides scientific ones. Reiner Schurmann in his *Heidegger on Being and Action; from principle to anarchy* (Bloomington: Indiana University Press, 1987), in the notes to his section 24 on alethea (truth), cites a leading follower of Heidegger who reads him as I do and presents a more sane reading. There is always a way to read an obscure text as sane, of course; all one need do is ignore or reinterpret the obscure, especially when the unpleasant is put obscurely. Embarrassingly for his disciples, Heidegger at times expressed his unpleasant ideas in crisp language.

12. See Isaac Todhunter's superb *A History of the Mathematical Theory of Probability* (Bronx: Chelsea, 1965), quoted by Ilkka Niiniluoto, *Truthlikeness* (Dordrecht: Reidel, 1987), p. 163.

13. See my *Towards an Historiography of Science* (the Hague: Mouton, 1963, 1967), pp. 56–57 and 112; see also my *Science and Society* (Dordrecht: Reidel, 1981), p. 336.

14. The literature on the theory of rationality is swelling. See my "Theories of Rationality" in *Rationality: The Critical View*, J. Agassi and I. C. Jarvie, eds. (Dordrecht: Reidel, 1987), pp. 249–63.

15. See Hermann von Helmholtz' "Faraday Lecture," in *The Faraday Lectures* (London Chemical Society, 1928), p. 133: "It is against the rules of probability that the train of thought which has led to . . . series of surprising and unexpected discov-

eries . . . should be without firm . . . foundations of truth. . . ." This is known in the literature as Fisher's likelihood measure: a hypothesis is likely if it renders improbable observation reports probable.

16. See E. T. Whittaker, *A History of Theories of the Aether. . . .*, and my *Faraday as a Natural Philosopher*, p. 110–16.

17. It has scarcely been noticed that the special theory of relativity is incomplete in the sense that there are different methods and different targets of approximation between it and other theories; I know of no writer on this except for Mario Bunge. See his *Philosophy of Physics*, pp. 184–85. On the whole, his discussion of inter-theoretic correspondence (chapter 9) is illuminating and substantially above the platitudes and popular errors perpetuated both by physicists and by philosophers.

18. See William Whewell, *On the Philosophy of Discovery* (London: J. W. Parker and Son, 1860), Appendix H, "On Hegel's Criticism of Newton's *Principia.*" See also my *Towards an Historiography of Science*, p. 112. For more details, see William Shea, "The Young Hegel's Quest for a Philosophy of Science or Pitting Kepler Against Newton," in *Scientific Philosophy Today: Essays in Honor of Mario Bunge*, J. Agassi and R. S. Cohen, eds. (Dordrecht: Reidel, 1982), pp. 381–97.

19. See my *Towards and Historiography of Science*, pp. 2, 79, 84–85, and 88–89. See also I. B. Cohen, "Newton's attribution of the first two laws of motion to Galileo," in *Atti del Symposium Internazionale di Storia, Metodologia, Logica e Filosofia della Scienza "Galileo nella Storia e nella Filosofia della Scienza,"* Collection des Travaux de l'Academie Internationale d'Histoire des Sciences, no. 16, Vinci (Florence): Gruppo Italiano di Storia dell Scienza, 1967, pp. xxv–xxliv: "Galileo seems to have put forward the view, rather explicitly expressed in his book on the sunspots, his *Dialogue* . . . that such inertial motion could be and was uniform circular motion. Only near or at the surface of the earth did he conceive of a true linear motion, and even that was apt to be a small arc of a very large circle." All this cautious circumlocution could not save Cohen from error: "a true linear motion" is never "apt to be [on] a small arc" not matter how "very large" the circle is.

20. See my *Towards an Historiography of Science*, p. 88.

21. The opening paragraphs of Kelvin and Tait, *Elements of Natural Philosophy* (Cambridge: Cambridge University Press, 1894), seem to me to say this rather shamefacedly. See my *Faraday as a Natural Philosopher*, p. 194. This point was lost when Cartesianism lost its appeal after the Einsteinian revolution, especially as Mach condemned Kelvin and Tait for their unorthodox Newtonianism; see his *Mechanics*, p. 245. Maxwell's unreadiness to revise Newtonian gravity is reported in Whittaker's *History*. . . . This, of course, is what made it so urgent to find the aether drift. It is a constant cause of embarrassment that historians of science and even textbook writers stress the significance of the aether drift or at least of the Michelson-Morely experiment, yet they do not even attempt to explain its importance or ask why it was considered important at the time. Nor can they explain why the idea of a field in empty space is so strange sounding that the majority of 19th-century physicists thought it was utterly out of the question. I discuss this point at length in *Faraday as a Natural Philosopher*.

22. Gerald Holton's valuable essay on Einstein's program, citing in note 2 above, has no explicit reference to Einstein's approximationism or even to his transcending Newtonianism. It does make the point, however, about Einstein's daring and ambition, and illustrates it in Einstein's earliest writings which were till then neglected.

23. See Karl Popper, "Three views concerning human knowledge," reprinted in his *Conjectures and Refutations* (New York: Basic Books, 1963. See also Einstein's letter to him published in Popper's *Logic of Scientific Discovery* (New York: Basic Books, 1959).

7

Newton, Nonlinearity, and Determinism

Frank Durham and Robert D. Purrington

"The complexity . . . is staggering, and I do not even attempt to outline it."

Henri Poincaré (1892)

1. BACKGROUND

In the last decade or so the previously rather esoteric science of nonlinear dynamics has suddenly become very fashionable. One reason for this is that classical nonlinear systems have been shown to have interesting and in many cases quite unexpected properties. Another is that such systems are not merely common, but actually the rule rather than the exception. While this had been known for a long time, it had been largely ignored because of the difficulty of treating such systems in detail. What has permitted the transition from the status of something of a curiosity to almost a fad is primarily the development of high speed digital computers. This is because few of these problems can be solved by analytical techniques. As we will be able only to suggest here, the study of nonlinear systems has had impact in fields as diverse as celestial mechanics and ecology, meteorology and cosmology.

One especially interesting property of some of these systems, those that are *strongly* nonlinear, is a seemingly random behavior exhibited by completely deterministic systems, known colloquially as "chaos." Another characteristic feature of such systems is an extreme sensitivity to initial conditions: systems with initial conditions differing by an arbitrarily small amount show outcomes, or final states, which diverge exponentially from each other. Naturally these developments raise interesting and profound questions concerning the deterministic, mechanical worldview we have inherited from the seventeenth century and require a reassessment of virtually every facet of Newtonian dynamics. The issues raised are both historical and philosophical; in the former area, it will be interesting to examine Newton's choice of problems to be treated in the *Principia* to see how he avoided, to the extent that he did, problems which are intrinsically nonlinear.

It has long been popular to speak of the Newtonian worldview as mechanistic and deterministic; Dijksterhus, Lovejoy, and Nicholson have described the replacement of the organic, hierarchical world-view by a mechanistic one in the seventeenth and eighteenth centuries.[1] To a considerable extent, the nineteenth century was the battleground between opposing camps of mechanism and vitalism (or other philosophical positions antithetical to the mechanical world view); it happens, also, that it was in the late nineteenth century that a particular branch of classical mechanics, nonlinear dynamics, had its origin. Its founders were mostly mathematicians, but by the turn of the century it was clear to physicists that many systems of physical interest were described by nonlinear differential equations and that most soluable problems were only linear approximations to the real descriptions.

It was also clear to these pioneers, Henri Poincaré foremost among them, that the problems in physics which are *linear*, whose solutions are provably stable, and can be combined or superimposed in simple ways, are special and rather few in number. Yet this understanding failed to work its way into the actual practice of physics in any extensive fashion. One looks in vain for references to the relevance of nonlinear dynamics in physics textbooks in the first three decades of this century.[2] The situation in mathematics is another matter. There are a variety of reasons for this, not least the computational intractibility of most nonlinear problems. The digital computer has removed this obstacle. Another reason for the relatively low profile of nonlinear dynamics no doubt has been the ascendancy of quantum mechanics, a strictly linear theory of the microscopic world. In any event, there is a wide awareness now of the existence of questions originating in nonlinear dynamics—questions that have fundamental implications for the history and philosophy of science.

Many of these questions are of such breadth as to make it inappropriate to pursue them here. Foremost among these perhaps is whether there needs to be a rereading of much of the history of dynamics in the light of new insights about old problems, although we must, to be sure, take care to avoid seeing the history of physics solely from the perspective of this century. We will content ourselves here with a brief look at the nonlinear problems that occur in the *Principia*.

A closely related problem is the unraveling of the complex and still somewhat obscure history of nonlinear dynamics itself. Since it involves the history of mathematics as well as of some the most subtle facets of classical physics, it will not be an easy task. Such questions as priority and influence have still to be sorted out. Since the origins of the story can be traced back to Newton himself, and since the overtaking of classical dynamics by the quantum mechanics is directly implicated in the incoherence of the field throughout most of this century, the definitive history when it comes will make interesting reading.

There are, not surprisingly, links between this new way of looking at macroscopic physics and longstanding philosophical questions such as determinism. Besides the direct implications of macroscopic causal randomness for the understanding of the scientific program, there is again the possibility of the need to reappraise at least the terms of some arguments that were thought settled. The rise of quantum mechanics notwithstanding, the examples and logic of Newtonian physics have for three hundred years been widely influential. What impact might a changed point of view about stability and predictability in the physical sciences have? We will examine this question below. There are, furthermore, new questions about the role of linearity as metaphor in the social sciences; we will outline a few such questions briefly.

Finally, we shall look at the problems Newton chose to treat in the *Principia*, with the new insights about nonlinear systems in mind, to understand why he chose them, where that is possible, and to consider how his success may have depended on his choice of problems.

Many of these questions have yet to receive the careful attention they deserve. Our goal is to at least sharpen a discussion of some of them. Because of the recent appearance of a number of reviews, both technical and popular, it should not be necessary to provide a full elaboration of the terminology of nonlinear dynamics, nor to discuss the major results.[3] Therefore, we will discuss only those results which are relevant to the questions at hand, especially the impact of these developments on our understanding of what Newton accomplished in the *Principia*.

2. STRONGLY NONLINEAR SYSTEMS AND NEWTONIAN PHYSICS

The modern field of nonlinear dynamics developed from several areas that initially had little overlap: Newtonian mathematical dynamics in the particular form called Hamiltonian dynamics, mathematical studies of nonequilibrium fluid flow, and the phenomenology of nonlinear systems of one or a few dimensions. Only recently have experiments been integrated into this mix of theory, and it is only to a degree that a fundamentally unified discipline can be said to have emerged.

The problem of stability for dynamical systems has a distinguished history in mathematical physics, as befits its status as the keystone of the Newtonian account of the world. It goes back to the *Principia*, where with great insight Newton was able to treat aspects of the solar system as a series of few-body problems. Nineteenth century mathematical techniques associated with William Hamilton built on the analysis of the great French mathematicians to tie dynamics to mathematical methods of subtlety and generality. Hamiltonian dynamics reached an astonishing plateau with Poincaré's demonstration that most conservative dynamical systems of more than two bodies are unstable under most conditions.[4] Although the Hamiltonian method was essential to the development of the quantum theory, the specific question of stability moved to the background as attention was focussed on the successes of the new quantum mechanics. But Kolmogorov and others took up the mathematics of conservative systems to produce elegant and surprising conclusions along somewhat similar lines as Poincaré's: namely, that many classes of such systems quickly lose all information about virtually any initial state.[5] The extension of these methods to the more general problem of nonconservative (dissipative) systems while maintaining full mathematical rigor has proved to be an enormously difficult problem. Nevertheless, many of the insights of modern nonlinear dynamics can be tied to this important body of work.

On a parallel front the study of nonlinear mathematical models with at least phenomenological connections to physical systems was proceeding. These models were relatively simple in form, but because nonlinear, were tedious or even effectively impossible to compute except for special cases. The development of fast digital computers changed this as early as the 1960s. Fundamental insights were gained through "mathematical experiments," direct numerical solutions repeated for many initial conditions. Many apparently disparate examples have been developed in this burgeoning field, some of which were seen to be related to traditional problems in dynamics, especially problems relating to the onset of turbulence in fluids.

Thus within the last fifteen years or so nonlinear dynamics has grown to include physical and life scientists, applied and pure mathematicians, engineers, information theorists, and others. While it may be too soon to say that the field has reached maturity in a Kuhnian sense, a considerable body of common terminology has developed. We will try to summarize relevant results from this field as transparently as possible. In particular we want to introduce some terminology that might pertain to systems that Newton himself considered, or that is relevant to one or another aspect of determinism.

We begin with one-dimensional Newtonian systems. Some of these are as simple as a single point mass subject to a few simple forces. The damped simple pendulum (with a constant gravitational force) and subject to a periodic but spatially uniform driving force is the prototype of such systems. Other one-dimensional models also have a frictional term and driving force. It is of course not difficult to set up experiments, whether mechanical or electronic, that incorporate these elements. With numerical techniques, computation of the complex motions of these systems is straightforward, but realistic experiments have appeared rather recently.

Dynamical equations for continuous physical systems form our next category. Newton's introduction of hydrodynamics in the *Principia* defined the first of a class of continuum problems in physics, problems for which some sort of averaging process had to be added to Newton's laws of motion. This process leads in plausible ways to partial differential equations such as those that describe fluid flow, heat flow, and diffusion. Such semi-phenomenological systems comprise a large part of the theory used in macroscopic physics and chemistry; much of engineering practice is founded on them as well. Nonlinear for most systems, they were from the beginning impossible to compute exactly by analytical techniques. Approximations at every level are required, and it is the solutions in such cases that are part of nonlinear dynamics. The traditional approach had been to determine equilibrium states; concentration on such solutions may hide the nonlinear character of the equations.

Because the partial differential equations describing physical models yield to analytical methods only for restricted classes of problems, processes by which the equations may be linearized have been important. Linearizing allows solution of the problem in certain special cases, or for certain values of the variables, but obscures its fundamentally nonlinear character. An approximate process of expanding the partial differential equations as a series of ordinary differential equations in many variables, and truncating the series to obtain a computable model has been used. E. N. Lorenz is credited with finding, in 1963, the first strongly nonlinear solutions ever clearly demonstrated, for a truncated system of three ordinary differential equations representing fluid flow under temperature gradient (as a model of the Earth's atmosphere).[6] Nonequilibrium chemical systems with catalysis are mathematically similar and can be compared to experiment.[7]

Virtually all stable strongly nonlinear physical systems are *dissipative systems*, systems with energy sources and sinks.[8] Such systems must damp or decay, unless there are sources or driving terms. The damping term can prevent instability and the driving term can prevent motion from being damped out altogether.

The phenomenological models that at first glance are farthest from Newton's equations are called iterative maps. These are equations in which some simple algebraic operation, the mapping, (involving squaring, adding a constant term, and the like), is performed on an arbitrarily chosen number to produce a new value, which is again operated on, and so on indefinitely to produce a sequence. Such equations, including difference equations, have been around for a long time, but were tedious to compute before the advent of digital computers. As with truncated systems of differential equations, some iterative mappings have applications—in biology and especially in ecology, as well as in the physical sciences—and others were treated simply for their mathematical interest.[9] Iterative maps as it turns out are tied to fundamental results in several fields of mathematics, including number theory and complex analysis, and to newly emerging subfields such as symbolic dynamics. To a considerable degree the mathematics of iterative mapping has provided the arena for the generalization and refinement of the varieties of nonlinear "dynamical" systems.

2.1 PHASE DIAGRAMS; CHAOS AND FRACTALS

A word about phase diagrams is necessary at this point, because they provide a common language to be used for discussions of continuous functions and for the number sequences of iterative mappings. A phase diagram at its simplest is a plane that displays values of two variables that are correlated. In physics this can be position and velocity (for a particle), or velocity and acceleration (for a fixed point in a fluid); or it can be successive positions (or velocities, etc.) separated by regular time intervals (fig. 7.1). The last of these correlations, which displays a sample of the continuously varying parameters of a physical system, is Poincaré's, and it provides a meeting ground for physically derived dynamics and iterative maps. The series of values obtained from the Poincaré map can be compared directly to the sequence of numbers obtained without intervening continuous evolution from an iterative mapping. The patterns themselves immediately indicate the type of solution—the "flow" of the system—and can be subjected to quantitative analysis derived from Hamiltonian theory. The latter connection is not surprising, since Poincaré's phase space sampling concept is a technique from Hamiltonian dynamics.

Whether the variables are positions, velocities, or pure numbers, if their successive values wander toward arbitrarily large magnitudes the solution is unstable. Solutions that never exceed some finite number are stable and are said to form an attractor on the phase diagram. A given system, whether computed or measured, may have have solutions of a quite different type for different values of the parameters (as opposed to the variables) of the system. (For one-dimensional Newtonian systems the strengths of the damping, driving, and restoring forces are parameters and the position and velocity are variables.) An equilibrium solution (for which a given variable tends to a fixed value) will approach a single point on the Poincaré phase plot. Periodic solutions, which always exist for the most-studied model systems, can be made to appear on the phase plot as one or more points (fig. 7.2). For continuous variables this requires a judicious choice of the sampling interval. Quasiperiodic solutions, if they exist, involve incommensurate periods (fig. 7.3) and will appear as closed curves. Stable solutions which are discontinuous and nonperiodic on the phase plot are called "strange

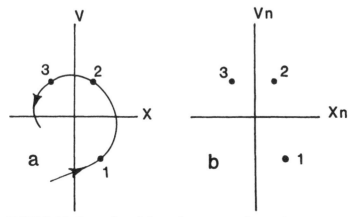

FIGURE 7.1. Examples of phase diagrams. In fig. 1a the trajectory of a particle is displayed by plotting the velocity v that the particle has at position x as time elapses. The labelled points correspond to fixed time intervals. If only the sampled points v_n and x_n are plotted, as in fig. 1b, the diagram is called a Poincaré plot. The Poincaré plot can show regularities not apparent on continous phase plots. See figs. 2 and 3.

attractors" and represent essentially new results of recent studies (fig. 7.4).

The strange attractors are also known as "fractal attractors," exhibiting a connection between the phase plots and fractals, mathematical entities that are intermediate among the usual geometrical constructs of point, line (or curve), surface, and volume.[10] The fractal construction is assigned a nonintegral "dimension" that corresponds roughly to the fraction of the space it fills. Within the relevant portion of the phase plane a strange attractor can range upward from dimension zero—a finite number of points in the plane— toward dimension two, which would correspond to every point in the plane belonging to the set.

In addition to not being limited to lines or areas, the strange attractors are "filled in" for any dynamical system in an aperiodic sequence that becomes essentially random. It is this feature that elicited the term "chaos," which was

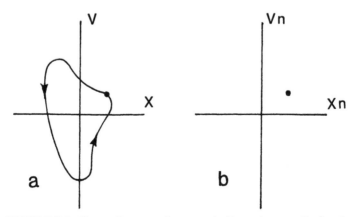

FIGURE 7.2. Phase diagrams for a periodic trajectory. In fig. 2a the trajectory is a closed path and the motion is periodic. When the time interval for sampling is equal to the period of the motion, the particle will always have returned to the same point on the phase diagram. On the corresponding Poincaré plot, fig 2b, the presence of only the one point (populated arbitrarily often) is diagnostic of periodic motion.

then adopted to refer broadly to systems capable of manifesting strange attractors.

Together these two properties reflect an exquisite sensitivity to the choice of starting point for the computation. If only one strange attractor exists within a given range of variables (positions, velocities, etc.) any starting point will produce a solution that traces out the strange attractor. An adjacent starting point will also map the attractor, but after some number of steps (sometimes one!) the sequence starting with the second initial point is uncorrelated with that from the first. Each sequence meets reasonable tests for randomness. Some systems, for example some with two stable equilibrium points, demonstrate competing attractors. In such cases the infinitely dense set of points (on the phase plot) not belonging to one attractor can include points that belong to the other attractor. For "fractal basin boundaries," as the case of competing attractors is called, chang-

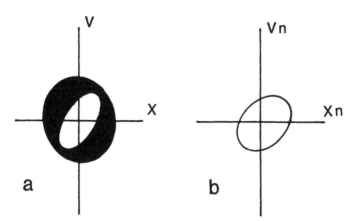

FIGURE 7.3. Phase diagrams for quasiperiodic behavior. Here the motion is bounded, but is not simply periodic. In an ordinary phase diagram, fig. 3a, the motion completely fills a region near the origin. The Poincaré plot for a properly chosen sampling interval reveals that the motion is quasiperiodic: in this case there are two incommensurate periods. For more complicated quasiperiodic motions the Poincaré plot can be filled in solidly as well.

ing the initial conditions arbitrarily little can result in a shift from one attractor to another.[11] Typically the transition from simple to chaotic behavior occurs as a system parameter is varied through some critical value. One way such a transition may occur is through a "crisis," an abrupt change from ordinary (periodic, say) solutions to the aperiodic solutions associated with strange attractors. In this case the transition to aperiodicity is signaled only by the occurrence, very close to the critical parameter value, of "intermittency," bursts of what look like noise but are the onset of fractal

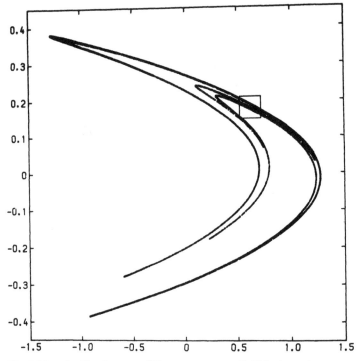

FIGURE 7.4. The famous "Hénon attractor." This iterative map corresponds logically to a Poincaré plot, although what is plotted is not successive samples of continuous motion but successive pairs of values obtained from a second order difference equation. On a large scale, as in fig. 7.4a, the plot appears to be a bounded curve,

behavior. Between bursts the solution will be quite regular (fig. 7.5).

An "intermittently periodic" sequence is of course not periodic. Lorentz pointed out that for a nonlinear system the solution *itself* can never demonstrate stability: intermittency may be present even in the longest finite run of regularity. As Lorentz says, for a solution sufficiently near the aperiodic region, "its very stability is one of its transient properties, which tends to die out as time progresses." Only experience with the system—empirical or fundamental

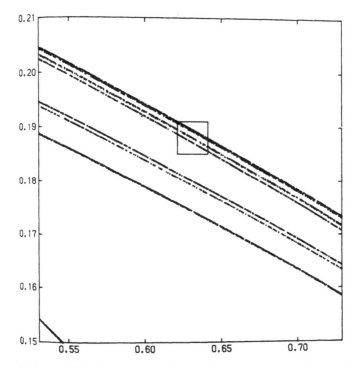

which would imply quasiperiodicity. Closer examination (fig. 7.4b) begins to indicate that the "curve" is a fractal structure. The "lines" are multiple on any scale, as Hénon showed. Expansion of successively smaller portions (as in the small box) always reveals the same degree of complexity.

knowledge of the correspondence between parameter values and types of solution—can settle the question of stability.

An important general insight about the behavior of nonlinear systems is also obtained by considering an alternative kind of transition to the strange attractor. Typically, a periodic solution is replaced, as the external parameter is varied, by successively more complex but still periodic or regular solutions. The transition from one class of solution to the next is called a bifurcation. The Russian physicist Lev Landau proposed in 1944 that fluid turbulence occurs by an

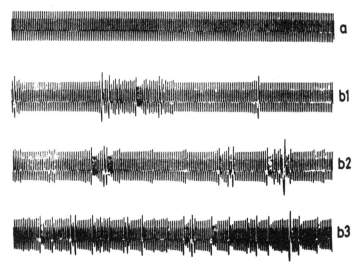

FIGURE 7.5. Examples of intermittent behavior. In this diagram, reproduced from Pomeau and Manville (reprinted in Bao-Lin, *Chaos*, ref. 6.), solutions to coupled differential equations are shown for four different but neighboring values of the control parameters. Each solution is for a *constant* set of control parameters, but only the top one appears to be periodic. The other three appear to be interrupted by intervals of "noise," but there is no noise in the equations (nor in the calculation). Such behavior calls into question, for strongly nonlinear systems, the use of the observed behavior to infer periodicity. Even plot *a* might be chaotic on a sufficiently long time scale.

accumulation of bifurcations, successively introducing more disorder.[12] Although that remarkable speculation proved not to be entirely correct, bifurcations have a special role for continuous physical systems. When a physical system encounters a bifurcation, no prediction can be made as to which "choice" will be made and how the new solution will be related in detail to the prior simpler solution. And repeating the same range of parameters with even slightly different initial conditions will change the solution. The effect is particularly dramatic when the bifurcation comprises a stable and an unstable branch. In that case, as the parameter is varied smoothly the particular solution either continues toward full complexity or diverges, again with arbitrary sensitivity to the initial conditions.

In the end, one still must raise the question of the universality of inferences to be drawn from nonlinear dynamics. To what extent can we expect a real nonlinear system to behave as the relatively simple model systems do? While the provisional answer seems to be in the affirmative, some caution is warranted, as we have already suggested. This is a difficult area where many results can only be demonstrated numerically, solid mathematical theorems are hard won, and conjecture and extrapolation is important.

2.2 DEGREES OF NONLINEARITY

It is a commonplace among physicists (although sometimes forgotten) that only a few mechanical systems can be adequately described as linear systems, that is, can be modeled using only linear terms in the equations of motion, i.e., Newton's second law. The small-amplitude pendulum, the two-body gravitational system, the ideal spring—these are a few examples, and the number is surprisingly small. On the other hand, many weakly nonlinear systems, which abound in nature and have straightforward mathematical description, can be treated perturbatively. A perturbatively stable solution is one which for which there exists an iterative correction to an unperturbed lower order approximation which is, in some sense, "small."

The rare but elegant linear systems have played a vital role in shaping the point of view of scientists. And in fact the simplicity of the linear system has over the past three centuries done much to encourage the presumption, by scientists and nonscientists, of the explanatory power of physical science. It is of course the broad relevance of the perturbative approach that actually underlies the assumption that much of the world can be treated as *approximately* linear.[13]

Strong nonlinearity, on the other hand, is a term which we will use to describe the situation in which the behavior of a system cannot be understood in terms of a perturbatively stable solution; only the direct solution of the nonlinear dynamical equations can provide useful information about the evolution of the system. Such computations cannot normally be carried out analytically, but require numerical techniques that must be implemented on digital computers. It is thus not surprising that the surge of interest in strongly nonlinear systems is coincident with the computer era, even though the mathematical foundations for much of the field were laid down much earlier. Of course we should keep in mind that the dynamical equations may describe the system only approximately, so that even an exact solution may be of limited generality.

There are further degrees of nonlinearity which will not concern us much here. Examples of such systems would include those for which essential properties are not even qualitatively reproduced by models. Such intractible problems have been given increasing attention by anti-reductionist "holistic" thinkers, who question or even deny the validity or applicability of conventional reductionist analysis of complex systems. The problem of mind is sometimes cited in this connection. The extreme manifestation of this view comes close to a denial of the relevance of the empirico-deductive approach of the "hard" sciences. Whether studies of strong nonlinearity can provide tools for treating such "essential" nonlinearity remains to be seen, but at the very least advances in nonlinear dynamics are helping to distinguish the truly intractable problem from the merely complex, or the straightforward but large-scale problem. Complex, highly organized systems need not be nonlinear in principle, but in fact are, given that the natural world is intrinsically nonlinear. As with all nonlinear systems, the question that arises is, what is the range of validity of the linear description? We have already seen that in many cases the range of validity is alarmingly small.

3. DETERMINISM, PREDICTION, AND OBSERVATION

Given the chaotic behavior of strongly nonlinear systems, one may be tempted to jump immediately to the conclusion that nonlinear dynamics sounds the death knell for determinism. The physicist might, if he were philosophically a bit subtle, assume that quantum mechanics had already eliminated discussion of determinism. That point requires lengthy analysis, but suffice it to say that quantum mechanics does not solve the problem, at least not trivially. It is convenient, in any case, to confine our attention to classical systems. Determinism for classical systems continues to be discussed seriously; it is neither a trivial, nor only a nineteenth-century question. In John Earman's phrase, determinism "leaves an indelible mark on every subject it touches."[14] Whether the question of predictability can be fully treated without invoking the framework of the quantum theory (or for that matter, special relativity; see Earman) is an open question , but there are reasons within the theory (including its essential linearity as theory) and without that lead us to believe that we can.

3.1 DETERMINISM

Let us begin by considering some of the terminology that has been developed to distinguish facets of what is broadly called determinism. Our interest is not in settling the issue, less still in attempting to force the reader to accept a single definition from among the many that are discussed in the

literature. Long before Laplace's famous statement of Newtonian determinism, examples from nature were used to make
concrete arguments for determinism or to counter such arguments. At the very least new insights from physics need
to be tested in the forum of such reasoning.

In this spirit we must reject the simplification that might
appear to follow from choosing a narrow meaning for determinism and hewing to it. Perhaps a dozen readily-distin-
guishable varieties of determinism can be found within the
literature. A useful starting point is to distinguish two broad
categories or families of constructs: determinism that is primarily logical, and determinism that is dependent in some
measure on physics. Let us examine these briefly.

Logical or "hard" determinism is propositional or syllogistic: all events are related to a cause, and therefore determined in the same way that from the truth of a proposition
its necessity follows. The physicist, accustomed by training
and practice to a sort of logical pragmatism, easily over-
looks the fundamental place that logical necessity has held
in philosophical tradition up to the present. Its assumed validity, grounded as it is in the syllogistic reasoning of Aristotle
and his successors, appears to transcend for many philosophers any details of definition for key terms or any question
of belief.

Michael White distinguishes "causal" determinism from
logical determinism as somewhat less propositional.[15] This
kind of determinism assumes that every occurrence in the
world has causes, which are not necessarily discoverable by
the methods of logic. If these causes are interpreted as physical, then a "block universe" is produced; the entire past
and future becomes an aspect of the single description that
is in the cause. Time in the block universe does not represent change, nor is it a variable, but merely represents the
limited perception of humans. Physical causal determinism
is close to hard Laplacianism, except that there is no requirement that the physical laws be Newtonian, or indeed
that they be known at all.[16]

Developments in physics of course cannot threaten logical determinism, which impinges on crucial questions of human freedom if taken to be part of the actual world, in which case it is a form of fatalism. A somewhat more comfortable "soft" determinism can be defined within logical determinism. "Soft" determinists (the phrase is due to William James, but the position goes back to antiquity) hold that free will and moral responsibility are not ruled out by universal logical determinism. White points out that this "compatibilist" position can be achieved by arguing that the chain of necessity is broken (or by some other related logic), or by simply denying the logical connection between cause and necessity.

A soft *causal* determinism can also be defined. Typically, advocates of this position would not take the block universe fully seriously, in particular rejecting the block future on the grounds that only the past and present are real. Milič Čapek has characterized this as the position of scientists (before the quantum mechanics) who accepted causal determinism but "were too empirically minded" to accept the unreality of time that the block universe requires.[17] Čapek argues that this represents a failure to understand that for hard causal determinism as for hard logical determinism effects "follow" causes not temporally but logically. In this connection it can be noted that soft causal determinism recovers human freedom as well.

Let us now turn to the status of philosophical questions that explicitly involve the tradition of physical law since the seventeenth century.

Any discussion of determinism and physics is likely to begin with a reference to the famous statement by Laplace that for "an intelligence which could know all the forces . . . and the states at an instant . . . nothing would be uncertain; the future, as the past would be present."[18] The interpretation of this Laplacian demon or spirit, and of the associated Laplacian determinism, deserves detailed analysis.[19] Let us, however, initially consider first a more obvious and less demanding category, which Karl Popper has called "prima facie" determinism and which has also been called ontologi-

cal determinism. [20] This consists in the observation that since
Newton the laws of nature have been expressed mathematically, and that a mathematical description conventionally
requires the specification of the variables used with absolute precision. On the face of it using mathematics then produces deterministic physics. In Stephen Toulmin's phrase,
this determinism is "the ghostly counterpart of our own
mathematics."[21]

Although they might not be familiar with the terminology, the majority of physicists who accept a modern version
of mechanism—the view that the universe is free of contin-
gency, that there are no influences that cannot be traced to
the quantitative laws of physics—in effect affirm both prima
facie determinism and causal determinism.

Three kinds of questions about prima facie determinism
present themselves. First, and most obvious, is the question
whether mathematics can be made to conform to nature exactly. "The unreasonable effectiveness of mathematics"[22]
persuades most scientists to be optimistic about this question. Whether mathematics maps nature *exactly* or only
"serves nearly enough" we will have more to say about below. Secondly, and of particular interest for the universality
of determinism, is the question whether mathematical cov-
ering laws originating in the physical sciences can be found
that are adequate to predict or even to describe all the phenomena of human experience. Dissenters from this view include the "emergentists," who are at least nominal determinists.[23] Finally, consider a related question: is mathematics
strong enough at its foundations to support the burden that
would be placed on it by the attempt to provide a complete
description of nature broadly construed? Popper and others
point to the work of Gödel and to paradoxes from specific
self-referential problems in order to discount the claims of
prima facie *universal* scientific determinism.

Laplacian determinism, which has had a long and complex interaction with the logical tradition of philosophy, depends on the structure of the solutions to the equations of
classical dynamics. The traditional reasoning runs as fol-

lows: the laws of classical dynamics can be cast as differential equations, whose solutions are exact except for constants that must be supplied as "initial conditions," which however can correspond to any given time. As Earman puts it, "for any times t_1 and t_2 and any allowed state at t_1, there is one and only one allowed state at t_2." The Laplacian demon, then, as the possessor of one "time slice" (space-like hypersurface) and the requisite equations, owns all prior and subsequent conditions as well. One can distinguish in the literature two versions of Laplacian determinism, this ideal mathematical version with initial conditions known exactly, and a less abstract version that only requires the initial conditions be known to within some bound. Popper calls the latter "scientific" determinism.

The distinction between hard Laplacian determinism and "scientific" determinism goes to the heart of the ambiguity about discussions of determinism when considered from the point of view of physics. The question here is that of the possession of *exact* initial conditions, a goal that experimental physics does not set for itself. This argument can be interpreted as placing hard Laplacianism more with "hard" causal determinism than with physical science.

Popper and other contemporary philosophers concur in this view of Laplacianism, and this provides the starting point for Popper's careful explication of "scientific" determinism, his version of the finite Laplacianism.[24] A "principle of accountability" is introduced which requires that the physics be capable of specifying the correspondence between the accuracy of initial conditions and the interval for which predictions can be made.[25]

A widely discussed premise is characterized as "programmatic" or "methodological" determinism, which has both logical and physical aspects. In its purest form programmatic determinism incorporates materialist reductionism, and assumes that because the world is deterministic and covered by physical laws, then the program of science consists in discovering more of the deterministic laws. To the extent that the ultimate laws would be consistent with

Laplacian determinism but would extend the language of physics to the entire range of experience, the outcome of the scientific program would be in effect to flesh out causal determinism. Other variations on programmatic determinism range downward in rigor, and include versions that countenance a broader range of laws than the materialist but leave open the question of observational omniscience. At another extreme, the range of programmatic determinism ends with a statement no stronger than a ratification of the scientific enterprise as capable of improving explanation—interpreted very loosely—beyond its present status. Sidney Hook has characterized a middle ground, in which non-dogmatic programmatic determinism is justified "on heuristic grounds" as more likely to produce useful results.[26]

Another form of physics-based determinism is "statistical" determinism. This apparent oxymoron involves the recovery of predictability in some sense, in the limit of large numbers of elements in the physical system. The obvious determinism of the gas law is cited as a reason for expecting macroscopic systems to behave in a predictable way. It is not so obvious that nonequilibrium systems can be expected to be statistically determinate, even leaving aside the difficult question of the role of statistics for quantum systems. Since we want to avoid a discussion of the full range of problems relating to probability, we only note here that in common with other "qualified" forms of determinism that have been described, statistical versions can be given precise meanings whose validity and implications can be explored.

One interesting variation on these themes originated with a modest suggestion by Bertrand Russell in 1912, who proposed to make the past, at least, determined by the device of listing all data (of unspecified type, but universal in some sense) up to a certain moment, then finding—or at least imagining—mathematical functions that correlate the data in unspecified ways.[27] No contextual meaning would be necessary for the functions. In effect Russell subordinates the role of physics in prima facie determinism: soft causal determinism is asserted to be trivially valid because mathematical functions exist. While the logic of Russell's argument seems generally to have been accepted, the relevance of his argument to programmatic determinism was challenged by Philip Frank and others.[28] More recently George N. Schlesinger has made the argument that Russell ignores the condition that physical laws must be simple in some sense, and that a mathematical function alone, without a defined context, does not make a physical law.[29] As we will discuss below, nonlinear dynamics has something to say about Russell's proposal.

3.2 DETERMINISM AND NONLINEARITY

Any attempt, such as Popper's, to construct boundaries in space and time for a "deterministic" physics could be taken as inconsistent with the premise of determinism. But the current interest in nonlinear dynamics makes the question of predictability a lively one, whether or not it is seen as originating in the history of the issue of Laplacian determinism. What indeed does predictability have to do with determinism? Some authorities, citing logical determinism as prior to science or irrelevant to human capacities, would say "nothing at all." John Earman, whose treatment of determinism is based on a rigorous mathematical approach, is particularly forceful in denying the connection between predictability and determinism.[30]

As the remarks of Čapek referred to above show, the block universe raises deep questions about the role and meaning of time. Further, as is often pointed out, an actual block universe not only makes prediction irrelevant, but also makes logic itself redundant, including the activity of discussing the block universe.[31] Within the tradition of physics this paradox has not been seen as reducing predictability to an irrelevancy. And, more to the point, prima facie determinism—the assumption that the full logic of mathematics is warranted in the pursuit of physical models—in effect equates predictability with the existence of solutions for physically relevant mathematical systems.[32] It is not necessary to go on to questions of observability or measurability to see that the exotic solutions, i.e., chaos, found within nonlinear dynamics speak to questions of predictability, nor is it hard to see that chaos and nonlinear dynamics are relevant to the historical question of determinism.

Consider again Laplacian determinism, not at all dead in classical physics unless one allows singularities such as infinite velocities and infinite energies, as Earman shows. All participants in the historical dialogue have accepted the "tyranny" of the differential equation, and its solutions, as that was understood before this century. Ernst Cassirer, in railing against hard Laplacianism, was reacting to what he saw as the resurrection of Leibnizian causal determinism in 1872 by du Bois-Reymond. The Laplacian spirit, Cassirer argued, is omniscient and cannot correspond to any human agency.[33] But in this he (and many others with him) explicitly accepted the assumption that the solutions to differential equations can be continued from one time slice for an arbitrary interval without loss of precision. It is by no means clear that this is the case, as we will discuss below.

(Popper's comments on the implications of nonlinear dynamics for Laplacianism involve a small historical puzzle. He accepts the equation of hard Laplacianism with simple omniscience; this is an early volley in his all-out assault on determinism. Introducing finite or "scientific" determinism, he attacks it by many arguments, some of which we have alluded to. In *The Open Universe*, published in 1982, he invokes an argument from Hamiltonian instability that he traces to Hadamard (1898).[34] What makes this interesting is that *The Open Universe* actually dates from the 1950s when discussions of nonlinear dynamics and determinism were not common, while by 1980 the numerical results on chaos had drawn a wide audience.[35])

Strong nonlinearity alters the distinction that has been made between Laplacian determinism and Russell's proposal for posterior determinism. Russell's austere version of determinism defines the past as determined because it can always be represented through mathematical functions of

arbitrary complexity. Objections to this insight have emphasized the distinction between arbitrary mathematical
functions and functions that could qualify as physical laws.
These reservations, as well as Russell's original proposal, need
to be reconsidered in the light of results from nonlinear
dynamics.

The first point at which the new mathematics (for we are
at this point discussing a philosophical model that has little
to do with experimental physics) enters has to do with the
loss of any general correlation between the degree of complexity of the dynamical equations and that of the motions
that are determined by them. If Russell's arbitrary functions are used to fit a collection of data from, as he says,
"the material universe" within a given time interval, then
logically all those data would be the trajectories of Newtonian
dynamics (or some more advanced future dynamics) applied
to the objects in the world. The "curves" would not then be
physical laws, but would merely be the outcomes of the presumably satisfactory dynamical laws. Nonlinear dynamics
shows that even for relatively simple Newtonian systems,
motions of arbitrary complexity are obtained. (Actual physical systems are subject always to the limitations of continuity for motion under finite forces.) These motions will over
appropriate time intervals include some that are indistinguishable from random motions by any finite test.

The linear paradigms from dynamics, such as free fall and
the planetary orbits, or the simple pendulum, encourage
confusion on this point, since in those cases the equations
and the solutions are of comparable simplicity. Historical
examples of "laws" contribute to this ambiguity, as with
Kepler's Laws, which of course were not grounded in any
general scheme. Russell may have been unaware in 1915 of
Poincaré's contemporaneous studies in instability.

Russell's own objections have to do, not with complex
trajectories per se, but with a more fundamental philosophical reservation about the *continued* simplicity of dynamical
laws. As he writes, "We cannot say that *every* law which has
held hitherto must hold in the the future, because past facts

which obey one law will also obey other, hitherto indistinguishable but diverging in the future."[36] The simplicity imposed on "acceptable" physical laws, therefore, is not only that they are successful in accounting for observed motions while preserving some connection with a body of definitions about the constituent world. Russell reminds us that in addition the laws should not arbitrarily change tomorrow to an entirely new set of laws, themselves self-consistent and mathematically simple. So, as regards the problem of simplicity, nonlinear dynamics has the effect of undercutting a long-standing objection to functional determinism. Functions of arbitrary complexity (consistent with continuity for coordinate motions) are perfectly acceptable in Newtonian dynamics.

There is more to be said about Russell's functional determinism at a more fundamental level. The essential result from modern dynamical theory concerning random determinate motions is the relation between randomness and *finite* information about the trajectory. In terms of Russell's empirical fits to trajectories, the modern conclusion is that any finite number of measurements is insufficient to allow extrapolation of a chaotic trajectory beyond a characteristic time that depends on the dimensions and velocities of the subsystem (interacting objects) under observation. Given that for the "material universe" subsystems of very small dimensions are numerous (We do not need to consider atomic dimensions), and given that chaotic orbits are not rare, it follows that in the actual material world no extrapolation can be made from the entire past, concerning the state of the *whole* universe.

What are the implications of these considerations for Russell's posterior determinism? One is that the functions postulated represent, not "the past," but only that portion of the past for which data are available densely. While it is easy to suggest that nature could have been instrumented for any given interval, Popper's objection that such a contention is "metaphysical" is now placed on mathematically solid ground.

Russell's determinism is seen to be logically close to Laplacian determinism. In each case one can distinguish an idealized and a practical version, only one of which has much to do with the practice of physics. The strong version of either can be interpreted as being an aspect of logical or prima facie determinism. In any case infinite precision in measurement, not merely for some time-slice, but for an indefinitely extended time interval, is required if the future is to be determined in the sense of physics.

We have faced, if not fully dealt with, the distinction between determinism and predictability. Consideration of the problems posed by chaotic systems makes it clear that a further distinction, as between computability and observability, is involved. For the mathematical models that are prima facie deterministic, it is computability that is more fundamental. Let us look briefly at computability and its relationship to the vocabulary of determinism.

3.3 COMPUTABILITY, CHAOS AND PRIMA FACIE DETERMINISM

We recall that prima facie determinism consists in the assumption that mathematically derived states of natural systems are deterministic because mathematics produces well-defined solutions. We have emphasized that this is a valid point of view, at least for those discussions of determinism that rely on examples from mathematical physics, whether classical or quantal. Earman in particular considers a variety of mathematical models for which uniqueness of solutions is at issue, and discusses computability at length.

We have noted in section 2 some of the surprising properties of strongly nonlinear systems. A crucial feature in discussions of computability is the existence of "strong mixing," the exponentially rapid divergence of solutions that were initially close to each other. As far as can be determined, strong mixing is present in real physical systems—as would be expected prima facie. Perhaps the most conservative statement we can make about this situation is that important questions of computability are still open. The problem

is exacerbated by the finite capabilities of digital computers—the only practical or indeed conceivable medium
for study of many strongly nonlinear models. Thus on two
different supercomputers the "same" initial point on a strange
attractor quickly diverges as a result of differences in han-
dling round-off. On any digital computer (supercomputers
are only faster in this regard) a computation repeats exactly.
But in a sense this deterministic outcome occurs *because* of
round-off. Computers simply cannot handle the real number
continuum for strongly nonlinear behavior.

One might easily call this ansatz uncomputability, and
mathematicians are currently vigorously engaged in trying
to sort out this aspect of nonlinear dynamics; the situation
is by no means intractable. Nevertheless, the meaning of
computability for typical nonlinear solutions in physics has
at least for now become more abstract. Actual computability has been replaced by conjectures about in-principle computability. It is perhaps too early to speculate as to the implications for prima facie determinism of practical problems
with computability. To the extent that the determinism implied by *physically relevant* mathematical models must be
assumed without persuasive proof, not prima facie but
the weaker programmatic determinism—quantitative optimism for science—is involved.

Paradoxically, strong nonlinearity can be seen as consistent with programmatic determinism, simply because it is
a scientific advance: the range of phenomena that can be
understood quantitatively has been enlarged. The basis for
optimism is explicit in the discovery of macroscopic *determined* randomness, where it is simply the equations, the laws
that are determined. At the new commonsense level, strong
nonlinearity encourages the expectation that arbitrarily
complex physical systems will eventually be characterized
by nonlinear dynamical equations. The most complex motions are in themselves no barrier to modeling: randomness
is not a sufficient condition for acausality. A change in outlook is involved. From the former assumption that with infinitely careful measurements the exact future could be ob-

tained—Laplace's "metaphor," according to some—dynamics offers now a quantitative but incomplete specification of the bounds of future motion. And most importantly the new characterization is not cut off from observation of the most traditional kind.

Classical randomness, of course, has been discussed since Boltzmann and statistical mechanics. But that was microscopic lack of knowledge, which could have been overcome by Laplace's demon and even perhaps by Popper's superscientist. A deep question for a century in statistical mechanics has been whether and how true randomness could occur for classical atomic dynamics, beginning with Maxwell's kinetic theory. We will not pursue this subject, except to make two remarks. First, it is clear that the same arguments that sharpen the definition of randomness for macroscopic systems apply a fortiori to classical atoms, at least to interacting systems of atoms. Secondly, randomness as loss of information about nonlinear systems is not restricted to any particular scale. The role of chaotic time evolution has begun to be considered even in astrophysical cosmology.

With these qualifications, programmatic determinism can be pursued with renewed intensity. From a distance this renewed optimism will probably not be distinguishable from the scientific enthusiasm that has prevailed in one form or another since the acceptance of Newton's insights in the Enlightenment. But it is a somewhat more informed optimism. Obviously, the program of nonlinear dynamics is not without its conceptual and practical difficulties. How, for example, can an appropriate model be found for a truly complex system? Even with large computers and large budgets, mathematical compromises must be made, dynamical approximations accepted.

And—an old problem but cast in new terms—how can the appropriateness of a model calculation be judged? No longer asked to reproduce exact trajectories, the new dynamics can only match observation in broadly quantitative ways. This problem is not necessarily insurmountable for "reproduction" of an observed class of motions, such as the emission of plasma jets from active galaxies. The statistical character of prediction is for many laboratory experiments in nonlinearity no problem at all, since by controlling the parameters the experimenter can avoid instability. Many systems in nature, however, are not so cooperative. In order to dramatize the problem of appropriateness of models one need only to look at the application of nonlinear dynamics in ecology. Here the models are in general phenomenological rather than fundamental. That the systems are nonlinear is not in question, and progress is made if only the basics of deterministic aperiodicity, intermittency, and other results from nonlinear dynamics can be communicated.

Persuading the constituencies—the workers, legislators, and consumers whose lives are affected by policies aimed at taking the parameters of the nonlinear model out of dangerous waters—that the model is correct before the fact, is a programmatic problem of a high order. Even after the fact it may be difficult to determine whether the system has been subject to strong nonlinearity, or whether quasilinear response to varying parameters accounts for the "chaotic" behavior. The dramatic proliferation and subsequent subsidence of the Crown of Thorns starfish on particular parts of the Great Barrier Reef may be such an example. Observers cannot agree as to the appropriate model—even the appropriate parameters—for the phenomenon. As a result they cannot agree whether the population changes were in any sense strongly nonlinear. While such controversies can be seen as characteristic of less fundamental phenomenologies, the point for programmatic determinism is the availability of a new arsenal of quantitative methods for incorporating apparently stochastic time sequences into a rational description. This is the point we emphasized earlier for the physical sciences.

We have drawn attention to the difficult question of computability for strongly nonlinear systems. Whatever problems this may or may not pose for prima facie determinism, the most intractible problems for mathematics are always self-referential. Strong nonlinearity for physical systems involves self-reference to the extent that effects that should be negligible within the system may not be. This is the "butterfly effect," Lorenz's term for the ultimately large effect in weather forecasting of initially negligible differences between two initial states. If a model for the weather does not incorporate the effect of a butterfly, for example, as well other relevant factors, any computation with the model is bound eventually to diverge from the weather with butterfly included.

From another point of view this amounts to a dramatization of the intractibility of open systems, for which strict determinism is out of the question.[37] Programmatic determinism can turn this situation to advantage by examining in a *statistical* way the role of small intrusions from the world outside the mathematical model. Such unpredicted influences traditionally are called "noise." The question of the role of noise in the scheme of nonlinear dynamics is an interesting one about which much has been written. Noise comes to attention in nonlinear dynamics for at least two reasons. First, for fixed parameters the solutions are randomly distributed on a strange attractor, and so may look as if noise were present when it is not. Intermittence is a dramatic instance of the appearance of what looks like noise. In this context it suffices to point out that this behavior is entirely produced by nonlinear mathematics and has nothing to do with noise as defined above.

On the other hand the sensitivity of nonlinear solutions to initial conditions, as well as the existence of bifurcations, implies that if noise is present—as invariably it is for macroscopic physical systems—it may play a decisive part in determining at least the particular history for a given solution. Prigogine and others, who emphasize the universal role of bifurcations for dynamical systems, tend also to emphasize the role of the stochastic in the evolution of a nonlinear system.[38] Economists in particular appear to be drawn to this aspect of nonlinearity. Typically, the sensitivity of a nonlinear model to noise can be determined only by computer experimentation. The general points again are that fluctuations are not necessarily noise, and that the question can be addressed systematically.

There is not space to treat in detail a most fundamental area of attack on prima facie determinism, one that strong nonlinearity is at the heart of. From fairly early in the century some commentators have quite properly objected to the assumption that computation is congruent with observation. This point is easily obscured if quantum mechanics is brought into the discussion. Leon Brillouin among others disputed prima facie determinism for classical physics. Already in the 1950s Brillouin, drawing on the legacy of Poincaré, was asserting that a distinction must be made between measurement and mathematical prediction.[39] At its most elementary level Brillouin's position is that the mathematical continuum has properties that cannot be attributed to instruments. Measurements, as he points out, never produce real numbers, but only integers or rational fractions. Poincaré's results already implied many of the nonintuitive features of chaotic motion, although Brillouin bases his arguments solely on the Hamiltonian dynamics. Now, after a decade of intensive study of strange attractors and fractal basin boundaries, the paradoxes are evident for a wider range of problems.

This class of ontological problems for the application of mathematics may have no relevance for physics, which so far has rarely had to turn aside from even the most nonintuitive implications of mathematics. For example, even in quantum systems with arbitrarily small dimensions the number continuum is used without difficulty. Ten years ago, when the force of determinate randomness first struck home, speculations about abandoning the number continuum were frequent. Drawing on the parallels with discrete parameters in quantum mechanics, Prigogine drew up the outlines of a classical description that might have satisfied Brillouin's concern about the unmeasurable.[40] Joseph Ford has also discussed this point at some length, and indeed speculations such as these are a source of references to nonlinear dynamics as a potential "revolution."[41] But chaos studies have settled in as standard subtopics in many journals. Not mystery or paradox, but an "alternate intuition," in Jack Wisdom's phrase, has been the fruit of nonlinear dynamics.[42]

4. NEWTON AND NONLINEARITY

"it will be sufficient if we find the changes no greater than may arise from the causes"

Isaac Newton, *Principia*

One of the outcomes of developments in nonlinear dynamics is that many problems that were thought to have been solved satisfactorily must now be reconsidered. Another is that other classes of problems have been found to be accessible after having been cast aside as too complex for detailed treatment. Much of the history of dynamics since the seventeenth century could with profit be reread with the new dynamics in view. It will be particularly interesting to look again at the problems that Isaac Newton himself treated, to see his approach to, or avoidance of such nonlinear systems. Such systems are ubiquitous in nature, and Newton did not shirk difficult problems. But while this is clearly a line of investigation that is worthy of pursuing, we content ourselves for now with a brief examination of the *Principia*. Even the cursory survey of the *Principia* with the new dynamical results in mind, which we make here, can reveal something of Newton's way of doing physics.

4.1 NONLINEARITY IN THE *PRINCIPIA*

It goes without saying that in the *Principia* nothing is included of mathematical solutions that imply strong nonlinearity. This does not mean that there was no examination of systems for which such solutions *can* occur. Yet the *Principia* is hardly a textbook of dynamics. Throughout Newton's life his *public* uses of the techniques of analysis were conservative or reactionary, sometimes to the point of being cryptic. This is evidenced from his earliest years by his reticence to share his methods of fluxions. Despite the proliferation of proofs, especially in book 1, the *Principia* from the beginning clearly reflects Newton's goals as well as his method: the demonstration of the system of the frame of the world; the demonstration of the universality of his laws; and the intimidation of potential critics of his results or his methods. In one way or another every derivation advances a broader argument about the physics, often to be taken up in book 3. Consequently we must look for instances of potential strong nonlinearity primarily among Newton's successes in an essentially focused treatment, not in a systematic exposition of dynamics. Many of the dynamical systems examined are not capable of exhibiting strong nonlinearity. Others are capable of such behavior but were treated only from a perspective that makes strong nonlinearity irrelevant. Newton does mention nonlinear systems that typically can behave chaotically. He does not always treat them, as we shall see, in the dismissive way that—from his example—served virtually to the present to distinguish the merely complex circumstance from the interesting problem.[43]

Above all there is celestial mechanics, especially the intractable problem of the moon which later came to be seen as the paradigm of the successful perturbative application of Newton's laws. Newton's choices of nonlinear celestial dynamics problems mostly demonstrate his talent for devising techniques of approximation. Newton's inventiveness is astounding, and his choices were influential not only in showing the way to development of more general methods; equally interesting is his ability to recognize the solvable aspect of a complex problem in celestial dynamics. Much of what he does as regards approximate solutions is narrowly directed at the problem at hand. This is further proof of his genius, for as Poincaré was to show, celestial dynamics is fraught with problems that cannot be treated by successive approximation. Clear indications of possible chaotic motions within the solar system are now beginning to be described in the literature. Such motions have always been present.

The least problematical nonlinear systems treated in the *Principia* are the highly damped one-body problems and the one-body central force problems.[44] One-body problems without a time dependent driving force will not show chaotic behavior unless damping is absent and an unstable equilibrium point is encountered in the motion. (Even then virtually all the solutions are analytical.) None of Newton's examples meets both of these criteria. Damped projectile motion is not a candidate for chaos unless degrees of freedom involving the shape of the projectile are taken into account; Newton treats point masses or centers of mass. One-dimension damped motions, such the pendulum in water, similarly hold no surprises.

Nevertheless, among Newton's one-dimension examples the best candidate *systems* would be pendulums, of which Newton considers a variety. A pendulum can show chaotic behavior when given enough energy to swing entirely around the support point. Again, to be of real interest the pendulum should be driven by a variable external energy source. Recently computer studies of the simple pendulum have appeared in such prestigious journals as *Physical Review Letters*. Such a rotor had no place in Newton's arguments, which characteristically use pendulums as means to various ends as diverse as understanding the resistance of fluid media or determining the shape of the earth.[45]

Similarly, the few instances of central forces in addition to the inverse square law are introduced either to advance the argument for the gravitation law, or to argue against Cartesian vortices. None of these cases contains an unstable equilibrium, and all are solved without ceremony. In sum, no potential chaos lurks among the one-body motions actually discussed in the *Principia*.

A substantial part of book 2 of the *Principia* is given over to fluids.[46] Newton's treatment of fluid systems goes beyond the consideration of objects moving through fluids to the invention of hydrodynamics, where the contiguous massive volumes in the liquid or gas not only respond to forces but also form the fluid, and the characterization of sound as a fluid disturbance.

One class of fluid motions, turbulence and incipient turbulence, has been the most studied type of strongly nonlinear motion. Many kinds of fluid systems are not chaotic, however, and Newton treats some of these at length. The instances of undriven pendulums in resisting fluids, which Newton analyzes and subjects to experiments, cannot be chaotic, as we saw above. The case of a U-tube with oscillating fluid level is mathematically similar to a damped pendulum; without a driving force nothing untoward happens.

Sound is introduced, quite remarkably given the context, as a fluid dynamical phenomenon in section 8 of book 2. The discussion however soon focuses essentially on the question of the velocity of sound, and to a delicate if somewhat dodgy attempt to show that the theory correctly gives the observed value.[47] As it turns out sound is one of the more resistant phenomena with regard to exhibiting strong nonlinearity. "Acoustical turbulence" has been observed recently at quite high pressure levels, and has been shown to have the earmarks of "chaos," but the demonstration is a tour de force of experimentation.[48]

Newton goes on to treat two kinds of hydrodynamical problems that are excellent candidates for nonlinearity under the right circumstances. These are flow from an orifice, and fluid in rotating containers. Among the earliest experimental demonstrations of chaos was the demonstration by the nonlinear dynamics group at the University of California at Santa Cruz that dripping faucets produce time sequences that include the onset of chaos, strange attractors, and the like.[49] For steady flow from an orifice or through a tube (Poiseuille flow) both laminar and turbulent regimes are seen, and intermediate kinds of strong nonlinearity. But Newton's interest in flow from an orifice is in the phenomenology of viscosity—a part of his campaign to demolish the Cartesian vortices—and the observations he details do not trace the flow rate over the range of values that might include bifurcations. Similarly, while Newton's introduction of bodies rotating in fluids is not so far logically from the Couette flow between *concentric* rotating cylinders, his eye is on the dreaded vortices, and his examples are mathematical.[50]

4.2 CHAOS IN THE *PRINCIPIA*

In the *Principia* Newton introduces at least two systems that evidence strongly nonlinear behavior in nature: comet tails and tides. Characteristically, he does not shirk the task of bringing within the net of his system phenomena that confounded the best minds of the seventeenth century. But he defines the terms of the discussion.

Of comet tails Newton writes that they must be "nothing else but a very fine vapor, which the head or nucleus of the comet emits by its heat."[51] In a wide-ranging exposition of semi-quantitative physics, he justifies this inference by arguments from the inverse square law for solar heat and from his knowledge of optics. He compares the comet tail to the rising smoke of heated bodies in order to account for the direction and curvature of the tail, and to account for the inverse relationship sometimes found between the length of the tail and the brightness of the head. Although he attributes to Kepler the (essentially correct) speculation that "their direction towards the parts opposite to the sun [is due] to the action of the rays of light" [p. 491] on the matter of the tail, he pursues the chimney analogy even though for the comet nothing is present that qualifies as a chimney, unless it is the atmosphere of the sun. Smoke rises due to its being mixed with air "rarified by heat" when "its specific gravity is diminished." Just so the longest tails develop when "the comets themselves are plunged into the denser and therefore heavier parts of the sun's atmosphere." This atmosphere, which is also referred to as "the matter of the aether," is somehow heated and rarified by the matter of the tail, so that a chimney effect occurs [p. 492]. Elsewhere in the discussion, however, the extreme tenuousness of the tail is derived and used to provide yet "another argument proving the celestial spaces to be free, and without resistance. [p. 491]"

The modern reader looking for chaos could wish that Newton had carried his analogy between comet tails and smoke columns even further, to a consideration of turbulence. But despite the evident role of turbulence in the motion of smoke, Newton rejects the opportunity to discuss such features of comet tails. "I add nothing," he writes, "concerning the sudden uncertain agitation of the tails of comets, and their irregular figures, which authors sometimes describe, because they may arise from the mutations of our air, and the motions of our clouds, in part obscuring those tails; or, perhaps, from parts of the *Via Lactea*, which might have been confounded with and mistaken for parts of the tails of the comets they passed by. [p. 489]"

The ease with which this question of observer reliability is let pass is in sharp contrast with his examination of the opinions about the cause of comet tails a few pages earlier.[52] There he calls on his own experience with comets both with telescopes and with the naked eye, in order to refute the argument that comet tails arise *entirely* from atmospheric effects. No contradiction is involved in Newton's casual treatment of turbulence. Commentators on strong nonlinearity have often described the gestalt involved, which almost amounts to a change from never seeing chaos to seeing it everywhere.

For all its evident inconsistencies, there is hardly a topic in the entire *Principia* that illustrates more cogently Newton's ability to construct an argument from incomplete data than this treatment of comet tails that is appended to proposition 41 (problem 21) of book 3. Newton argues persuasively, that is to say quantitatively, that comet tails are diffuse, that they are subject to gravity, that they are composed of particles moving from the nucleus to the end of the tail and on into space. He speculates, less persuasively, that the oceans are replenished from the material evaporated from the solid bodies of comets. In the process he takes even this most ephemeral feature of a comet out of the realm of mystery and brings it down to the mundane. The tail is indeed rather like the exhaust of a chimney, even if he has the detailed mechanism wrong. And of course one does not follow the properties of the mundane to unreasonable lengths, to the point of the unexplainable. Turbulence for smoke is beneath consideration; turbulence for comet tails is dismissed from sight. Such lessons in style were not lost on the physics community that grew from the fertile ground that was the *Principia*.

The problem of the tides would be a subtle one even if the earth were covered everywhere with deep ocean. Newton in effect begins his discussion of tides with this simplified model. The gravitational attraction of the moon and sun raises tidal bulges, whose motions are driven by the changing positions of those bodies referred to the earth's surface. Here already is a candidate for strongly nonlinear motion: the ocean as driven oscillator, damped by the viscosity of seawater. Newton comments that "the two luminaries excite two motions, which will not appear distinctly, but between them will arise one mixed motion compounded out of both." [p. 415] And that "the motions . . . suffer some alteration from that force of reciprocation, which the waters, being once moved, retain a little while by their *vis insita*." [p. 417]

Add the nonlinear effects of shallow bottoms and of constraining boundaries and channels, and aperiodicity—chaotic motion in the broad sense—is the norm. Newton discusses these complicating factors at some length and recognizes their role. He goes so far as to describe aperiodicity (though he would not have used the word) occasioned when "it may happen that the tide may be propagated from the ocean through different channels towards the same port, and may pass quicker through some channels than through others; in which case the same tide, divided into two or more succeeding one another, may compound new motions of different kinds." Multiple periods or stagnation could occur at different times, with phase reversals, and he describes these phenomena. "An example of all which Dr. Halley has given us, from the observations of seamen in the port of Batsham, in the kingdom of Tunquin" where two inlets mix the tides from the China Sea and the Indian Ocean. [p. 418]" All this is remarkably close to a characterization of strong nonlinearity.

Newton's effort to prove that his quantitative treatment of the tides is correct reverts to the consideration of *average* tides on the English coast at various times in the lunar month. Here he goes beyond the approximate treatment that accounts for the spring and neap tides and for longer-term modulations he associates with the extremes of the orbital distances, and compares his computed ratio of 9:5 (for spring and neap tides at the equinoxes) to Samuel Sturmy's observations of 45:25 feet for this ratio [p. 450]. Westfall has discussed the evident overstatement of the precision for both the calculations and the observed tides, as an example of Newton's quantitative polemic.[53] The computation of tides is intricately involved with Newton's unhappy attempt to determine the precession of the equinoxes to high accuracy. Granting this, Newton's discussion of tides, taken as a whole, nonetheless represents a valid way to deal with strongly nonlinear motions. He recognizes the crucial parameters— the driving forces of moon and sun, the inertia and viscosity of the sea, and most importantly the depth and shape of channels—and sorts out their relative contributions. To a degree he separates the solutions into essentially linear and strongly nonlinear types. On the whole he recognizes the impossibility of exact solutions for real boundary conditions, although he cannot resist putting his stamp ("9:5") on the tides of the river Avon.

4.3 NONLINEAR CELESTIAL DYNAMICS AND THE *PRINCIPIA*

We have reserved consideration of the question of strong nonlinearity in the solar system. Here is the primary subject of the *Principia*, and the solar system is obviously nonlinear. Indeed so many nonanalytical problems are there in plain view, beginning with the moon, that they motivated Newton's most extended efforts and led to much of what has followed in dynamics. We must limit our examination of this complex set of problems to an overview and a few general remarks.

The first question here is how it should happen that strong nonlinearity can keep a low profile in the solar system. The answer lies in the near absence of damping for a primarily gravitational system. The solar system is a many-body problem for Newtonian dynamics, and in general many-body problems—"many" meaning more than *two*—have no exact solutions and can include strongly nonlinear trajectories. However the two-body problem in Newtonian gravitation—the sun and one planet, all other objects ignored—is a linear problem. Consequently, to the extent that the motions of the planets are completely dominated by the sun, "the frame of the system of the world" is an almost linear one.[54] One of Newton's key insights was that only for the inverse square law do the axes of the planetary ellipse remain stationary. By treating directly the motions of the apsides he could not only establish the role of perturbation but also treat it quantitatively.

For the planets as they are—with neighboring planets and in the case of some a massive satellite—as well as for the satellites themselves and for debris floating between and across planetary orbits, exact analytical solutions will not always serve nearly enough. A reasonable approximation, as Newton saw, is to treat the effect of the weaker of the other two bodies as a correction to the Keplerian trajectory produced by the dominant body, which is the sun for a planet, or the planet for a satellite. There is no guarantee that this procedure will work, but it is the only way to proceed with pen and paper. If still a fourth body is important, the same perturbative approach is used again. This is only a three-body approximation in the sense that after the dominant body, other bodies are treated one at a time. In any case we are referring to the celestial bodies treated as mass points; the problem of rotation almost always can considered separately.

The typical outcome for *undamped* gravitational interaction among three or more bodies is instability: the ejection or coalesence of some of the bodies. According to the standard model of solar system formation, enormous selection has taken place since the early, dirty proto-system was formed. After five billion years only the most stable components remain; strong nonlinearity played a role in this process. Nevertheless, scope exists for the observation of instability and complexity within the solar system, and even on the human time scale.

Recollection that the solar system is truly a many-body system implies that chaotic motions are a certainty for some objects in the solar system, at some times. In recent years calculations have taken this insight beyond the level of speculation. Jack Wisdom and others have developed numerical techniques for treating many-body problems within the solar system. These calculations have demonstrated chaotic motions among asteroids; in particular, intermittent and dramatic changes in the eccentricity of asteroids are predicted that account for the puzzling "Kirkwood gaps" in the asteroid belt. The "Digital Orrery," a dedicated parallel processor, has within the past few years extended the span of plausibly accurate computation of planetary orbits from about one hundred thousand years to several hundred million years. Concentrating on the orbit of Pluto, the most eccentric of planets, Gerald Jay Sussman and Wisdom have apparently demonstrated that Pluto's motion meets all the tests for chaotic motion.[55] In particular the computed orbit shows exponential divergence of orbits computed from nearly the same initial conditions. A third example of computed strong nonlinearity from this research program is the chaotic tumbling for Saturn's satellite Hyperion obtained with a model that uses the nonspherical shape found by Voyager 2.[56]

These calculations, as impressive as they are, are far from the last word on the solar system. And as always with strongly nonlinear systems the results have a generic flavor in that the motions produced cannot be the actual motions. It will be interesting to follow the development of these nonperturbative treatments of the solar system to see what further surprises may lurk. Meanwhile, they provide quantitative support for conjectures about strong nonlinearity in the solar system.

Comets are the only class of objects to include gravitationally unstable trajectories that were accessible to observational astronomy in Newton's day. Conventionally, the comet originates in the Öort cloud far outside the planets, and its inward trip is triggered by a three-body interaction. This destabilizing event is invisible from Earth. While in the inner solar system the comet's motion may be transformed to a less eccentric orbit by a close planetary encounter. In that case—one thinks of Halley's comet—the comet is for a time *dynamically* quite like an asteroid: small, and continually subject to the perturbing influence of the planets. But comets were easily visible to Newton and his contemporaries, while asteroids were unknown.

The question is not whether comets can display chaotic orbits; Wisdom's calculations for asteroids already demonstrate this possibility, since the nongravitational forces on a comet are essentially irrelevant to classifying its orbital motions, although above the level of the "butterfly effect." It is, rather, whether any of the comets Newton actually discusses were (or are) in strongly nonlinear orbits. In fact much of the detailed material on comets in the *Principia* is drawn from Halley's work, which includes additional results, but we can make our point without pursuing this.

The only candidate available to Newton for cometary chaos is Halley's comet itself. In the *Principia*, periods are assigned only to the comet of 1682 (Halley's) and to the much more dramatic comet of 1680—which D.W. Hughes argues should be known as "Newton's comet," since the latter's observations of the comet figure crucially in his later treatment of comets in the *Principia*.[57] The comet of 1680 is tentatively assigned a period of 575 years based on records of comets in 1106, 531, and "in the month of September after Julius Caesar was killed." [p. 479] Since Halley was able to find a suitable fit to a calculated 575-year period, the periodicity is taken to be established. Later calculations showed this procedure to be incorrect; the comet had not previously been seen during historical times.[58]

Halley's insight that the comet of 1682 might be periodic rested on the similarity of its inferred trajectory near perihelion to that of the (less precisely observed) 1607 comet, together with other retrograde comets which appeared at appropriately earlier times. While the sensational returns of the comet since 1682 have produced the public misapprehension that the comet is indeed strictly periodic, Newton knew from Halley that successive apparitions of the comet differ by a year or more from the "nominal" 75-year "period."

Characteristically, Newton minimizes this deviation from Keplerian simplicity. "But," he wrote, "because of the great number of comets . . . and of the slowness of their motions in the aphelions, they will, by their mutual gravitations, disturb each other; . . . Therefore we are not to expect that the same comet will return exactly in the same orbit, and in the same periodic times: it will be sufficient if we find the changes are no greater than may arise from the causes just spoken of." [p. 502] Newton correctly identifies the sensitivity of the comet's orbit at aphelion, when a slight change in velocity will have greatest effect on the subsequent orbit. For comet Halley and other medium and short-period comets, the perturbing effect of the Jovian planets is much the larger cause, since comets are not as massive as Newton thought.

One or a few deviations from periodicity form a slender thread on which to hang a discussion of chaotic motion for comets; as we discuss below, settling the question observationally for a particular comet may be difficult indeed. The problem is that perturbative calculations may serve well enough to compute deviations from periodicity for a given orbit (and may be updated at each return), without, however, predicting the long-term motion in any but an average way. The success of the Newtonian program is assured "if we find the changes no greater than may arise from the cause" attributable from gravitation. But the essential limitation to the earlier Newtonian goal—not Newton's goal, obviously—of exact prediction of celestial motions already lurks in the description of Halley's comet.

One other area of nonlinear celestial dynamics Newton treated involves the rate of rotation of planets, satellites, or asteroids. The appearance of chaotic variations in rotation, which can be thought of as being driven by tidal forces, have as we noted been computed for a model of the Saturian moon Hyperion. Strongly nonlinear rotations are absent for spherical bodies, and can be prevented by damping due to internal friction for planets and large satellites. Even today direct observation of such motions would very difficult for asteroids or small satellites.

One object qualitatively meets all the conditions for observable chaoticity: the moon. Yet at least over historical time the moon's motions can be accounted for through the methods of iterative perturbations that Newton used unsystematically and the French mathematicians advanced in the century that followed. In the history of dynamics, Newton's frustration at the intractibility of the lunar motions is cast against the dramatic and crucial question of the completeness of the Newtonian dynamics for describing celestial motions, and by extension its relevance to the broader question of the mathematicization of nature.

If it is somewhat fortuitous that the problem of the Moon did yield, largely, to Newton's persistence, one might speculate what the impact of the *Principia* might have been had the moon's motion proved quite impossible to describe. The qualities of lunar dynamics—its close coupling to Earth and the lack of rigidity of both bodies—that frustrated and infuriated Newton, make perturbative treatment problematical over very long intervals. And these same intractable features make the moon an uninviting candidate for direct nonperturbative calculations. No treatment of the moon as a strongly nonlinear system have been reported to the present time, to our knowledge. Perhaps when computational techniques are developed to handle such problems, a chaotic past or future for the moon will be laid bare. If more accurate computations—or measurements—should reveal that Newton's efforts at a perturbative treatment of the problem of the moon were (in one respect or another) doomed to be frustrated, it would only be seen as one more success for Newton. On such contingencies does the history of science turn.

4.4 GENERALIZATIONS ABOUT CHAOS AND THE SOLAR SYSTEM

The calculations by Wisdom and others for asteroids (valid also for comets) demonstrate the possibility of spectacular jumps in orbital eccentricity. A second, less dramatic kind of bounded chaotic motion would involve aperiodic (but not secular) variations in the orbit. Here a very interesting point for the language of strong nonlinearity occurs. To represent orbital variations on a phase plot would amount to plotting one point for each (average) orbital period. Over many orbits the distribution would tell the story: the nature of the attractor or attractors would develop. The time interval for the development of randomness—the loss of predictability—would however be related to a combination of the periods of the comet and of one or more planets, potentially a very long time. Consequently, the deviations of the chaotic comet from a simple ellipse could be tracked "microscopically" by perturbative methods for a comparably long time. Furthermore, in the spirit of Popper's "scientific" determinism, if the error in a *single* orbit is not great then the orbit is determined—nearly enough. A prediction for some reasonable interval might not in general involve those encounters (bifurcations) that confound perturbative calculations. Given the short history of human civilization, observed comets with large eccentricities fall into this middle ground. Whether multiply periodic, quasiperiodic, or aperiodic, they are likely to be seen at first as "periodic," and then as perturbatively "almost periodic"; and finally as "potentially chaotic." To find the true motion from observations is beyond human capacity; to compute the true—as opposed to characteristic—motions is not an appropriate goal. In this instance strong nonlinearity restores a certain residue of mystery to the comets that Newton so thoroughly demystified.

5. EPILOGUE

In the end, a review of Newton's choices of nonlinear systems and how to deal with them reveals much about the design of the *Principia* and about Newton's approach to the problem he set himself—nothing less than conversion of philosophy to a mathematical enterprise. In hindsight it appears that Newton's working from the phenomena to the explanation, or offering a mathematical polemic (as against vortices) helped to keep him on solid ground. With notable exceptions, much of his work with complex systems consists more of thrusts, or skirmishes, than of the kind of siege that he waged against gravitation—the inverse square law and projectile motion.

Perhaps the most remarkable thing about the universe is that it *appears* to be linear. Is this because we have been indoctrinated by three hundred years of dynamics, or it is because human beings experience the universe in domains of time and frequency which admit of an approximately linear description? Is quasilinearity to become a hypothesis that we no longer need? Newton the eschatologist would have had no trouble with that question.

NOTES

1. E. J. Dijksterhus, *Mechanization of the World Picture*, C. Dikshoorn, trans. (Oxford: Clarendon Press, 1961). Marjorie H. Nicholson, *The Breaking of the Circle* (Evanston, Ill.: Northwestern University Press, 1950). A. O. Lovejoy, *The Great Chain of Being* (Cambridge: Harvard University Press, 1961).

2. The 1925 textbook by Max Born on "atomic mechanics" or the old quantum theory discusses problems of stability for conservative systems. Published in English as *The Mechanics of the Atom* (London: Bell, 1927).

3. See for example C. Grebogi, E. Ott, J. Yorke, "Chaos, Strange Attractors, and Fractal Basin Boundaries in Nonlinear Dynamics," *Science*, 238 (1987):632; or D. R. Hofstadter, "Strange Attractors: Mathematical Patterns Delicately Poised between Order and Chaos," *Science*, 225 (1981):22. James Gleick's journalistic tour de force, *Chaos: Making a New Science*, (New York: Viking Penguin, 1987), is much the best known of the popular treatments.

4. A "conservative" system is in effect one for which mechanical energy is constant. Poincaré proved that for most dynamical systems at most one integral—the total energy—exists.

5. A comprehensive but technical review is G. M. Zaslavskii, *Chaos in Dynamic Systems* (New York: Harwood Publishers, 1985). Many of the issues of complexity for Hamiltonian systems are discussed accessibly in Ilya Prigogine's *From Being to Becoming* (San Francisco: W. H. Freeman, 1980).

6. E. N. Lorenz, "Deterministic Nonperiodic Flow," *J. Atmos. Sci.*, (1963):130. Reprinted in Hao Bai-Lin, *Chaos* (Singapore: World Scientific, 1984).

7. Generic equations of this type have been studied, e.g., by O. E. Roessler, "Different Types of Chaos in Two Simple Systems," *Z. Naturforschung*, 31A (1976):1664.

8. The term "dissipative system" is used in a different but related sense in catastrophe theory. An accessible discussion of catastrophe theory is contained in Ivar Ekeland, *Mathematics and the Unexpected* (Chicago: University of Chicago Press, 1988).

9. Among the earliest discussions of biologically relevant models is Robert M. May, "Simple Mathematical Models with Very Complicated Dynamics," *Nature*, 261 (1976):459. May in 1976 was already urging that "the most important applications, however, may be pedagogical . . . [and] that people be introduced to [this subject] early in . . . mathematical education."

10. The number of dimensions is of course not limited to three.

11. The problems involved in sampling fractals with finite numbers have received much attention. The consensus is that the inferences given here are robust. See *Science*, 241 (1988):1162. But surprises may await.

12. L. Landau, *Akad. Nauk. Doklady*, 44 (1944):339. English translation reprinted in Hao Bai-Lin, *Chaos*. 1984.

13. Thus, consider the reports that only buildings of a certain height in downtown Mexico City were at greatest risk from the long-period vibrations originating from a distant earthquake. The image that is suggested, of a resonant response that depends on mass and linear dimension as for a tuning fork, is valid in approximating the behavior of a building before it begins to fail structurally, even though the response of a building to large amplitude shaking is not at all linear.

14. John Earman, *A Primer on Determinism* (Boston: D. Reidel, 1986), p. 249.

15. Michael White, *Agency and Integrality* (Boston: D. Reidel, 1985), p. 30.

16. White argues that one distinction between the two categories is that the logical version requires a level of omniscience that goes beyond "mere" knowledge of the future to an involvement of the omniscient being with the rules of logic.

17. Milič Čapek, "The Unreality and Indeterminacy of the Future in the Light of Contemporary Physics," in *Physics and the Ultimate Significance of Time*, David R. Griffin, ed. (Albany: State University of New York Press, 1986), pp. 297–308.

18. Pierre Simon de Laplace, *Philosophical Essay on Probability*, 1814.

19. N. David Mermin has provided a charming commentary on the infiltration of the philosopher's "ean," into the "ian" domain of physics, in the case of "Lagrangian" in "What's Wrong with this Lagrangean?" *Physics Today*, 41 (1988):9.

20. Karl Popper, *The Open Universe* (Totowa N.J.: Rowman and Littlefield), 1982.

21. Stephen Toulmin, *The Philosophy of Science* (New York: Harper, 1960), p. 162.

22. We take the phrase from Eugene P. Wigner's "The Unreasonable Effectiveness of Mathematics in the Natural Sciences," in his *Reflections and Symmetries* (Cambridge: MIT Press, 1970), 222–37. But see also Kant.

23. R. W. Sperry, *Phil. Sci.*, 53 (1986):265, and references therein.

24. Karl Popper, *The Open Universe*, 1982.

25. Popper's goal was the demolition of determinism, so that his cannot be considered to be an objective approach to the question. Nevertheless, the discussion of limited knowledge as a factor in classical science is important and useful.

26. Sidney Hook, "Necessity, Indeterminism, and Sentimentalism," in *Determinism and Freedom in the Age of Modern Science*, Sidney Hook, ed (New York: New York University Press, 1958), p. 168.

27. Bertrand Russell, "On the Notion of a Cause," reprinted in *Mysticism and Logic and other Essays* (London: George Allen and Unwin. 1917), pp. 180–9.

28. Philip Frank, *Philosophy of Science* (Englewood Cliffs N.J.: Prentice-Hall, 1957).

29. George N. Schlesinger, "Is Determinism a Vacuous Doctrine?" *Br. J. Phil. Sci.*, 38 (1987):339–46.

30. John Earman, *A Primer on Determinism*, p. 7.

31. See Popper's discussion of this argument, especially of J. B. S. Haldane's version of it, in *The Open Universe*, section 24.

32. The reader may dissent from this definition; we do not intend to imply that this is Earman's meaning of predictability.

33. Ernst Cassirer, *Determinism and Indeterminism in Modern Physics* (New Haven: Yale University Press, 1956), part I, pp. 3–28.

34. Karl Popper, *The Open Universe*, section 14.

35. See the foreword to *The Open Universe* by W. W. Bartley, III, the general editor of "The Postscript to the Logic of Scientific Discovery," pp. xi–xvi.

36. Bertrand Russell, *Mysticism and Logic*, p. 4.

37. Ivar Ekeland discusses this question in an accessible way in *Mathematics and the Unexpected* (p. 64 ff), where he refers to calculations by M. Berry on the dramatic gravitational effect of unobserved objects for strongly nonlinear systems.

38. Ilya Prigogine, *From Being to Becoming*, pp. 147–50; and references therein.

39. Leon Brillouin, *Scientific Uncertainty, and Information* (New York: Academic Press, 1960). This volume incorporates as well several of Brillouin's earlier journal articles.

40. Ilya Prigogine, *From Being to Becoming*, chap. 8.

41. Joseph Ford, "How Random is a Coin Toss?" *Physics Today* (1983):40.

42. The phrase is used by Jack Wisdom in his 1986 Urey Prize Lecture, published as "Chaotic Dynamics in the Solar System," *Icarus*, 72 (1987):241.

43. A fairly complete list of nonlinear systems in the third edition is as follows.

Nonlinear one-body and equivalent one-body motions
    Projectiles with resistance
    Nonlinear pendulums
    Central forces other than inverse square
Three body problems
    General considerations
    Motions of apsides
    Orbits of moons
    Cometary orbits
Rotations of nonspherical celestial bodies
Fluid systems
    Flow from an orifice
    Objects moving in fluids (per se)
    Oscillation in a U-tube
    Sound
    Rotating containers
    Tides
    Comet tails

44. For two bodies, and central forces such as gravitation, the motion of either body relative to the mutual center of mass corresponds to the "equivalent" one-body problem.

45. See Richard S. Westfall's essay in this volume.

46. We mean fluids treated explicitly. Motion with velocity-dependent resistance implicitly involves fluids; if these cases are included the role of fluids is even larger.

47. Richard S. Westfall, "Newton and the Fudge Factor," *Science*, 179 (1973):751–58.

48. Werner Lauterborn and Eckehart Cramer, "Subharmonic Route to Chaos Observed in Acoustics," *Phys. Rev. Let.*, 47 (1981):1445–48.

49. See, e.g., Harold Froeling, James P. Crutchfield, J. Doyne Farmer, Norman H. Packard, and Robert S. Shaw, "On Deter-

mining the Dimension of Chaotic Flows," *Physica*, D3 (1981):605–17.

50. The famous rotating pail described within the Definitions is closer to a realizable experiment, but would lack the control over parameters that makes Couette flow so rich in results.

51. Isaac Newton, *Mathematical Principles of Natural Philosophy*, 3rd edition, Andrew Motte, trans., New York: Daniel Adee, 1848, p. 486. Page numbers given after other *Principia* quotations refer to this edition.

52. P. 487. Of course Newton could not let pass so fundamental a question as the status of comets; the argument that comets are atmospheric and ephemeral had been advanced by Galileo himself, in *The Assayer*.

53. See Richard S. Westfall, *Never at Rest* (New York: Cambridge University Press, 1980), p. 736.

54. Westfall has pointed out that in the argument for design, a solar system with nearly perfect symmetry looks rather like a botched job. Similarly, with regard to Kepler's laws "nearly elliptical" is no law at all. See, in the latter connection, Joseph Agassi's essay in this volume.

55. Gerald Jay Sussman and Jack Wisdom, "Numerical Evidence That the Motion of Pluto is Chaotic," *Science*, 241 (1988):433–37.

56. Jack Wisdom, "Chaotic Dynamics in the Solar System."

57. D. W. Hughes, "The *Principia* and Comets," *Notes Rec. R. Soc. Lond.*, 42 (1988):53–74.

58. D. W. Hughes, "The *Principia* and Comets," pp. 64, 71.

8

Transcending Newton's Legacy

Henry P. Stapp

S cience influences our lives in many ways. That of technology is evident, but effects on social institutions, such
as church and government, can be equally important. Consider, for example, the apparent impact of Newton's idea of
"law" on the drafters of the U.S. Constitution, and ultimately on the rights and freedoms of its citizens.[1] More important, probably, than these examples, is the the influence
of science upon our idea of what we are; upon our idea of
our place in the universe, and our connection to the power
that forms it. For our aspirations and values spring, in the
end, from our idea of what we are; nothing is more important than the character of the ideas which motivate our
actions.

Science was transformed during the twentieth century by
three revolutionary developments: the special theory of relativity, the general theory of relativity, and quantum theory. These developments altered not only scientific practice
but also our ideas about the nature of science and the nature

of the world itself. I shall discuss here these three developments with attention to both their essential differences from classical Newtonian science, and their potential impact upon the human condition.

1. NEWTONIAN SCIENCE

"Newtonian science" must be distinguished from the full thought of Isaac Newton; it may be characterized by the following conditions:

1. Absolute Time and Absolute Space. Newton's starting point is the idea of a "true" time and a "true" space. Each is independent of anything external to it, and has an inherent quality of uniformity or homogeneity. These two "absolutes" are contrasted by Newton to their "relative" or "apparent" counterparts, which we can grasp through our senses, and can measure by means of clocks and rulers.
2. Local Ontology. Absolute space is conceived by Newton to be populated with small bodies or particles that move with the passage of absolute time.
3. Fixed Laws of Motion. The motions of the particles are governed by "laws." These laws allow the locations and velocities of all particles at *all* times to be determined by the locations and velocities of all particles at any *single* time. The world is thus *deterministic*; its condition at one time determines its condition for all time.

These features of Newtonian science give us a picture of the universe called the Mechanical Worldview. According to this view the universe consists of nothing but objectively existing particles moving through absolute space in the course of absolute time in a way completely determined by fixed laws of motion.

This picture of the world is mathematical: the objects are described mathematically, by numbers that give the locations and velocities of all the particles. Moreover, the laws that govern these numbers are mathematical. That Newton aspired to the creation of a mathematical picture of Nature is proclaimed by his title: *Mathematical Principles of Natural Philosophy*.

2. THREE PROBLEMS

Some difficulties with this picture of Nature were evident from the start. I mention three:

1. Action-at-a-distance
2. Creation
3. Freedom

2.1 THE PROBLEM OF ACTION-AT-A-DISTANCE

The centerpiece of Newton's science is the law of gravity. According to this law, every body in the universe acts instantaneously upon every other one, even though they be separated by astronomical distances. Newton's recognition of a problem with this idea is expressed clearly in his famous assertion: "That one body can act upon another at a distance through the vacuum without the mediation of anything else . . . is to me so great an absurdity that I believe no man, who has in philosophical matters a competent faculty of thinking, can ever fall into it."[2]

The ontology set forth in the *Principia* provides, however, nothing to mediate the force of gravity. Newton worked hard to find a carrier for gravity compatible with the available empirical evidence, much of which came from his own experiments. Finding in the end nothing that met his standards, he declared: *"hypotheses non fingo"*—"I frame no hypotheses."

Two contrasting attitudes toward physical theory can thus be found in Newton's thinking. One attitude reflects his basic overriding commitment to search for truth about Nature. This commitment is massively displayed by his extensive researches into alchemical and theological questions pertaining to the constitution of Nature, by his choice of title mentioned above, and by his careful attention, in the formulation of his principles, to philosophical and ontological details. The second attitude goes with his *"hypotheses non fingo."* This declaration entails that his theory, as it stood, must, strictly speaking, be construed not as an ontological description of Nature itself, but merely as a codification of connections between measurements. The theory must be viewed as a system of rules that describes how our observations hang together, not as a description of the underlying reality.

These two contrasting attitudes toward physical theories will be the focal point of my discussion of how Newton's ideas fared in the twentieth century. The issue concerns two views of the nature of physical theory. One view holds that basic physical theory ought to provide a description of the real stuff from which the universe is constructed—it should describe the ultimate things-in-themselves. The second view holds that physical theories should deal fundamentally with quantities that can be measured—that they should merely codify the structural features of the measurable phenomena.

2.2 THE PROBLEM OF CREATION

Given the precepts of Newtonian science, two questions concerning the problem of creation immediately arise.

1. What fixed the nature of the particles of the universe and their laws of interaction?
2. What fixed the initial locations and velocities of the particles?

Within the framework of Newtonian science, these questions are insoluble. Then, if one holds the first of the two views described above, that physical theory should describe the real world, the account provided by Newtonian science is deficient, for it requires something external to the physical world it describes: it needs something to set up the system and fix the undetermined parameters.

From the second point of view, which is that science should merely codify, not explain, this problem of creation might seem to be no problem at all. But the problem in this case is with the point of view itself, which tends to close off the pursuit of further knowledge. For today, within the framework of the quantum theory, physicists are examining theories that purport to answer the first of the questions raised above, just on the basis of self-consistency. Moreover, the second question is moving into science in connection with studies of the birth of the universe—the big bang. The question is, therefore, this: to what can science aspire? Can it cope with the problem of creation, or must it remain forever mute on this fundamental question?

2.3 THE PROBLEM OF FREEDOM

Beyond these question lies one of great immediacy to man. The mechanistic worldview proclaimed by Newtonian science, and "validated" by its technological success, insists that all free creative activity ceased with the birth of the universe. It tells us that we are now living in a mechanically fixed universe that grinds inexorably along a path preordained at the birth of the universe, held in place by immutable laws of nature. Thus any notion that, by our own efforts, we can act to bring into being one state of affairs rather than another is sheer illusion and fantasy. This dreary view is proclaimed in the name of science, and is backed by its authority. Banished, together with freedom, is any rational notion of human responsibility. For responsibility can be placed only where freedom lies, and according to the precepts of Newtonian science all freedom expired when the universe was born.

I shall return to examine these questions from the perspective of twentieth century science, but first an essential stepping stone from the ideas of Newton to those of twentieth century science must be introduced.

3. GALILEO AND LORENTZ

The laws of Newton have a simple consequence: given one possible universe, evolving in accordance with Newton's laws, it is possible to construct another simply by adding to the velocity of every particle in the universe any arbitrary common velocity. All separations between particles are left unchanged, and, according to Newton's laws, this shifted state of [affairs] will perpetuate itself through all time. This property is called Galilean invariance.

In 1873 James Clerk Maxwell proposed a theory of electric and magnetic forces that was wonderfully beautiful and marvelously successful. This theory did for electricity and magnetism what Newton had tried to do for gravity: it explained the forces between charged particles in terms of changes that propagate from point to neighboring point, thus abolishing the need, in electricity and magnetism, for action-at-a-distance. However, Maxwell's theory was characterized by a certain maximum speed of propagation, the speed of light in a vacuum. According to this theory, no charged particle could move faster than this maximum speed. Consequently the property of Galilean invariance was lost. However, Maxwell's theory offered a replacement, which involved this characteristic maximum speed. This new property, called Lorentz invariance, was to play a crucial role in what lay ahead.

4. THE ABSOLUTE VERSUS THE RELATIVE IN TWENTIETH CENTURY SCIENCE

4.1 THE SPECIAL THEORY OF RELATIVITY

According to Newton's idea of absolute time one can assert that if A and B are two events, each of negligible duration, then either A is earlier than B, or B is earlier than A, or they are simultaneous. The truth of any such assertion, that "A is earlier than B," for example, is absolute and does not depend on anything else.

Consider, however, two such events (in space and time) A and B, situated so that nothing can move from either event to the other without traveling faster than light. In this case one cannot determine by direct observation (say the observation of one event from the location of the other) which event occurs earlier than the other. One might expect that such a determination could be achieved by indirect means. Einstein, however, showed that if all phenomena in Nature enjoyed the Lorentz invariance property then it would be impossible *in principle* to determine from empirical data which of the two events occurred first.

The Lorentz invariance property seemed to hold universally (phenomena associated with gravity excepted, since Newton's theory of gravity needed to be reformulated along the lines of Maxwell's treatment of electromagnetism). Consequently, Newton's idea of absolute time seemed to bring into physical theory a property that *in principle* could have no correlate in observable phenomena. Einstein therefore proposed that physical theory be based not on absolute space and time as Newton had proposed, but rather upon a spacetime structure defined by idealized readings of clocks and measuring rods. The resulting theory is the special theory of relativity. Physicists quickly accepted this idea, which produced economy in notation and conception. Thus they replaced the absolutes of Newton by their relative counterparts.

4.2 QUANTUM THEORY

Quantum theory is another twentieth-century construct that makes measurement primary and fundamental. It carries the shift from absolute to relative even further than the special theory of relativity. For, according to the orthodox view of quantum theorists, not only must the underlying spacetime framework be understood in terms of results of possible measurements, but, in fact, the entire mathematical formalism of quantum theory must be interpreted merely as a tool for making predictions about results of measurement.

This view of quantum theory arose from its historical origin and its intrinsic form. But it is sustained by a reason far more compelling than mere "economy": every known ontology that is compatible with the phenomena, as codified by quantum theory, is "grotesque" in some way. Orthodox physicists, reluctant to embrace the grotesque, prefer to adopt a rational stance that separates the predictive mathematical formalism and the associated scientific practices from ontological speculations that lack empirical support.

5. CONVERSATIONS BETWEEN EINSTEIN AND HEISENBERG

Werner Heisenberg was the principal creator of the formalism of quantum theory. He has given an account of an interesting encounter with Einstein.[3] He prefaces this account with a brief description of the genesis of quantum theory; how he, reflecting upon Einstein's claim that a physical theory should contain only quantities that can be directly measured, and realizing that orbits of electrons inside atoms cannot be observed, was led to discover rules that directly connect various measurable quantities which arise in experiments performed on atomic systems, without making reference to unobservable orbits.

Early in 1926 Heisenberg described this new quantum theory at a symposium in Berlin attended by Einstein. Later, in private, Einstein objected to the feature that the atomic orbits were left out. For, he argued, the trajectories of electrons in cloud chambers can be observed, so it seems absurd to allow them there but not inside atoms. Heisenberg, citing the nonobservability of orbits inside atoms, pointed out that he was merely following the philosophy that Einstein himself had used. To this Einstein replied: "Perhaps I did use such a philosophy earlier, and even wrote it, but it is nonsense all the same." Heisenberg was "astonished": Einstein had reversed himself on the idea with which he had revolutionized physics!

To find the probable cause of this "astonishing" reversal it is necessary only to look at what Einstein had done between the 1905 creation of special relativity and the 1925 creation of quantum theory. The special theory holds, as mentioned earlier, only to the extent that the effects of gravity can be ignored. To pass to the general case it was necessary, as we have noted, to generalize Newton's theory of gravity.

Einstein undertook this task and in 1915 announced his general theory of relativity. Though this theory was a generalization of the special theory in many ways, it was fundamentally different. The focus was no longer on observers and the results of measurement. The theory was about a spacetime structure that exists by itself, governed by its own nature, without relation to anything external. It was about an "absolute" spacetime structure. Einstein was driven by demands for rational coherence and by a general principle of equivalence. He sent his work to Born saying that no argument in favor of the theory would be given, since once the theory was understood no such argument would be needed.

Einstein had in this work gone beyond *"hypothese non fingo."* He had succeeded in doing what Newton had been unable to do. He had discovered a mathematical description of something that *could* be regarded as Nature itself. The difficulty that defeated Newton, namely the action of gravity at a distance without any carrier, had been resolved by first combining Newton's absolute time and absolute space into an absolute spacetime, by relaxing Newton's demands for uniformity, and finally, by imposing his mathematical laws in the form of conditions on deviations from uniformity: the presence of matter was represented by departures from uniformity, by distortions of spacetime itself.

An important difference between Einstein's theory and Newton's is that in Newton's theory time and space are independent of each other, and both are independent of matter. This creates, at least in principle, the possibility of space with nothing in it: an empty arena.

The idea of empty space has puzzled philosophers since antiquity: how can anything be nothing? Thus Newton's predecessor Descartes takes extension, hence space, to be something that cannot exist without matter. Newton's con-

temporary Leibniz takes space to be merely a system of relations. Yet it remains puzzling that so much of the universe
can be (almost) empty if empty space is nothing at all.

Einstein's ontology gives a marvelous solution to this ancient puzzle. Instead of three intrinsically different things—
space, time, and matter—whose connection must, from a
logical point of view, be *ad hoc*, hence puzzling, we have
only one thing: inhomogeneous spacetime.

Considering the direction and achievements of Einstein's
general theory of relativity, one cannot be surprised that its
creator should regard the philosophy of the creator of the
special theory of relativity as "nonsense all the same."

The fate of Newton's two absolutes in the twentieth century is then this: the special theory of relativity replaced
them by their relative counterparts, but the general theory
resurrected them in a combined form that incorporates also
the third element of Newton's ontology, matter. But quantum theory represents a swing from the absolute back to the
relative. For according to the orthodox view, quantum theory must be viewed as a codification of connections between
measurable, or relative, quantities.

With this background in place, I now turn to the question
of the impact of twentieth century science upon our ideas
about Nature, and upon our ideas about ourselves.

6. IMPACT OF QUANTUM THEORY UPON
THE MECHANISTIC WORLDVIEW OF
NEWTONIAN SCIENCE

In general, quantum theory gives only statistical predictions. The question thus arises: does Nature itself have gen-
uinely stochastic or random elements? Neils Bohr stated the
orthodox position: that in practice, no matter how precisely
we prepare an atomic system, there is still a scatter in the
results of certain experiments. Quantum theory gives predictions with a matching irreducible scatter. Thus the statistical character of the theory matches the statistical character of the facts. To say more than this is empirically

unsupported speculation: quantum theory says nothing about determinism in Nature. Quantum theory successfully describes and predicts phenomena on the basis of a mathematical description of atoms. Can we conclude that the world is built of atoms?

If one looks at the mathematical representation of these atoms, one finds entities that must, according to the orthodox view, be interpreted *only* as parts of a computation of expectations pertaining to results of measurement. Thus the ontological foundation is shifted from the level of the atoms to the level of the devices that record these results, or perhaps even to the level of the observers who use these results to make computations. But the devices and observers are assumed to be built from atoms. So the ontological basis swings back to the atom, and so on.

These examples illustrate the difficulty in trying to draw ontological conclusions from a theory that must be interpreted merely as a tool for making predictions about connections between measurements.

7. QUANTUM THEORY AND REALITY

It is clear to everyone that we cannot pass with certainty from knowledge about the structure of phenomena to knowledge about the structure of the underlying reality. Accordingly, the orthodox interpretation of quantum theory tries to isolate, as far as possible, the mathematical formalism, and the scientific practice associated with it, from more speculative activities: it tries to separate "science" from "natural philosophy." Science is concerned with measurable quantities, and with theoretical structures that codify the observable and testable connections between them. Natural philosophy concerns the conclusions that might reasonably be drawn about the form of the underlying reality on the basis of the evidence provided by science.

The fact that Bohr and Heisenberg adhered to the view that the mathematical formalism of quantum theory should be viewed, strictly speaking, merely as a tool for making predictions pertaining to results of measurements in no way implies that they had no interest in the implications that quantum theory has in the realm of natural philosophy. In fact, each in his own way tried to draw from the data provided by quantum theory insights into the nature of the world that lies behind the phenomena.

8. HEISENBERG'S ONTOLOGY

Heisenberg in his book *Physics and Philosophy,* in the chapter on the Copenhagen interpretation actually sets forth an ontology.[4] He begins with the words "*If* we want to describe what happens in an atomic event." He then goes on to describe an ontology in which the actual world is formed by "actual events," which occur only at the level of the macroscopic devices. But the objective world contains also something else—"objective potentia." These "objective potentia" are objective tendencies for the actual events to occur. They are associated with the mathematical probabilities that occur in quantum theory.

This ontological substructure gives nothing testable. So it is not "science." But it gives us an informal way of "understanding" quantum theory. It gives us an idea of what is actually going on.

This ontology described by Heisenberg is not the only ontology compatible with the predictions of quantum theory. But it can be said to be the "most orthodox" ontology. Most quantum physicists probably think about quantum phenomena informally in these terms: the quantum probability function corresponds somehow to the *tendency* for the detector to register a particle, or the *tendency* for a grain in a photographic plate to register the absorption of a photon. The actual things occur only at the macroscopic level.

The reason it is interesting to consider ontologies suggested by the structure of phenomena as codified by quantum theory, and compatible with this structure, is that the conditions thus imposed on ontologies are so restrictive: there is no known ontology that is compatible with the conditions on phenomena imposed by quantum theory that is not "grotesque" in the minds of conservative thinkers. This means that quantum theory has shown us that the world is not at all like what we had previously imagined it to be. It is not at all like the idea of the world set forth in the mechanical worldview, formerly promulgated in the name of science, and still largely dominating the prevailing idea of what science tells us. So any curious person must naturally be led to ask: what idea of the world *is* compatible with the data provided by science?

9. WORLDVIEW ARISING FROM HEISENBERG'S ONTOLOGY

Heisenberg's ontology is the most orthodox, and in my opinion, the most reasonable, of the known ontologies that are compatible with the predictions of quantum theory. I shall now describe the principal features of the picture of Nature that arises from this quantum ontology.

9.1 THE WORLD IS NONLOCAL

Macroscopically separated parts of the universe are linked together in a way that involves strong faster-than-light connections that do not fall off with increasing spatial separation. This nonlocal aspect is the "grotesque" feature of this ontology that makes it unacceptable to conservative thought. However, one can show that under very general conditions this nonlocal aspect inheres in the statistical predictions of quantum theory themselves, and is hence almost unavoidable.[5]

9.2 CREATION IS DISTRIBUTED OVER ALL TIME

In the quantum ontology, the object potentia are represented by the quantum probability function. At each stage the quantum potentia give tendencies for the next actual event. The occurrence of this next actual event is represented by a "collapse" of the potentia to a new form. The interplay of the Heisenberg uncertainty relations and the

Heisenberg equations of motion is such that, even though each successive event effectively closes off certain possibilities, by making fixed and settled things that had formerly been unfixed, still, each event creates new potentialities and possibilities. Consequently, the process of fixing the unspecified degrees of freedom, which in classical physics occurs all at once, at the creation of the universe, is, in the quantum ontology, by virtue of its mathematical structure, a process that can never close off the possibility of its further action. Thus in the quantum ontology, the creative process, in which things formerly unfixed become fixed and settled, does not expire at the birth of the universe, but extends rather over all time.

9.3 TWO KINDS OF TIME

The quantum ontology defines time in two distinct senses. The first is "Einstein Time," which joins with space to form Einstein's spacetime. The second is "Process Time." The differences are as follows.

The "numbers" that appear in Newton's theory, and which describe the positions and velocities of the particles, are replaced in quantum theory by "operators" which evolve in accordance with Heisenberg's equations of motion. This evolution of the quantum operators represents evolution in Einstein time. It generates an association of operators with spacetime points: every spacetime point, from the infinite past to the infinite future, becomes associated with a *fixed* set of operators.

The spacetime structure just described is a structure of quantum operators. To obtain the potentia one must take these operators in conjunction with what is called the Heisenberg state vector, which does not depend on spacetime but rather refers to all of spacetime. It combines, however, with the operators associated with any spacetime point to produce numerical potentia associated with that spacetime point.

Each actual event corresponds to a "quantum jump" of the Heisenberg state vector to a new value. Thus each actual event induces at every spacetime point a sudden jump in the potentia associated with each point.

The sequence of quantum jumps defines a time that is different from Einstein Time, called Process Time. Evolution in process time generates change or evolution of the "actual," whereas evolution in Einstein time generates the evolution of the "potentia." Thus the deterministic laws of evolution are not binding on our future, for they determine the evolution of the potentialities, not the actual events themselves.

9.4 MEANING IN THE QUANTUM UNIVERSE

The creative process is represented in the quantum ontology by the sequence of jumps in the quantum potentia. These *potentiae* are objective tendencies, which tend to make the statistical predictions of quantum theory hold under appropriate conditions. But the question arises: what determines the actual course of events? That is, what determines, in a given actual instance, whether things will be fixed in one way rather than another? Heisenberg's ontology gives only a statistical condition on the choices of the actual events. Hence this ontology, as currently understood, may be incomplete.

At first it might seem that, in any case, the choice of what actually happens is either deterministically fixed by what has gone before, or has an element of true randomness or wildness. In either case, the ontology would appear to provide no possibility of a meaningful universe: either we would have simply a new determinism, which would render the universe just as mechanistic, and devoid of possible meaning, as the world of Newtonian science, or there would be an element of randomness, which could hardly add meaning. Thus we are apparently still trapped between the two horns, determinism and randomness, of the usual dilemma of the impossibility of a universe with meaning.

To have meaning, a choice must have intentionality; it must exist in conjunction with an image of the future that it acts to bring into being, or to block. Any choice that does not refer in this way to the future lacks meaning. In the Newtonian picture the future does not exist in the present, and hence it cannot enter into any present event, or choice. The future cannot be changed by any event or choice.

In the quantum ontology, the future does exist objectively in the actual present, albeit as potentia. Thus the future can enter into the present event. This event can, moreover, by altering the potentia for the future events, effectively block or bring into being a chosen state of affairs. In this sense a quantum event can, in principle, have intentionality and meaning.

The quantum ontology *allows* for meaning, but does not demand it. Indeed, insofar as it is complete, the quantum ontology described above does not provide meaning. For, in this case, the actual choices must be considered purely random, subject only to the probability structure. However, if a "principle of sufficient reason" holds, then this apparent randomness must be an expression of the incompleteness of that ontology. Then the quantum ontology, with its objective potentia defined over all spacetime, provides a natural basis for a deeper ontology in which every actual choice is meaningful in the sense that it actualizes an intention for organization that is emergent in that actual event itself. Each actual event then occurs in the context created by the prior (in process time) actual events.

9.5 MAN IN THE QUANTUM UNIVERSE

The role of man in the universe is tied to the mind-body problem. From the perspective of the quantum ontology, the brain is a macroscopic system similar to a measuring device. The function of the brain is to organize input and then to make a decision that initiates an appropriate action. According to the brain-device analogy, this decision is represented as a quantum jump. Just as in the case of a measuring device, this quantum jump is a macroscopic event: the whole brain, or some macroscopic part of it, is involved.

The problem of understanding, within the framework provided by classical physics, the connection between consciousness and the physics of the brain, is illustrated by the following passages cited by William James.[6]

"The passage from the physics of the brain to the corresponding facts of consciousness is unthinkable. Granted that a definite thought and a definite molecular action in the brain occur simultaneously; we do not possess the intellectual organ, nor apparently any rudiment of the organ, which would enable us to pass, by a process of reasoning, from one to the other." (Tyndall)

"Suppose it to have become quite clear that a shock in consciousness and a molecular action are the subjective and objective faces of the same thing; we continue utterly incapable of uniting the two, so as to conceive that reality of which they are opposite faces." (Spencer)

The problem here is that the classical conceptual frameworks for describing the physical and mental aspects of Nature contain no common, or even similar, elements. According to classical physics the physical aspects of Nature are completely reducible to motions of the separate molecules or other particles. But the mental aspects are described in terms of feelings. There is no logical connection between a feeling as such, and a motion of billions of separate molecules.

The quantum ontology has an analog of the classical motions of molecules moving in accordance with Newton's laws. It lies in the evolution of the corresponding quantum operators in accordance with Heisenberg's equations. However, the quantum ontology has something else as well, which has no counterpart or analog in classical physics: the actual event.

Within quantum ontology the conscious event and the physical event can be naturally understood as the psychological and physical faces of the very same thing, namely the event of selecting and initiating a course of action. On the psychological side there is the felt or conscious event of selecting and initiating this action, and on the physical side there is the physical collapse of the potentia that selects and initiates this action: the physical brain, as represented in quantum mechanics, collapses to a state in which the instructions that initiate the particular course of action are actualized. The connection between these two events is not an *ad hoc* and arbitrary identification of things as totally disparate as, on the one hand, a motion of billions of separate molecules, and on the other hand, a unified conscious event. It is, rather, the association and identification of the felt event with the physical event that represents, within the quantum ontology, exactly the event that is felt. In this way conscious events become special instances of the actual events that, according to the quantum ontology, form the fabric of the entire actual universe.

CONCLUSION

Quantum theory does not *entail* any specific ontology, and it is unreasonable to expect that it should. However, the "most orthodox," and, I believe, the most reasonable, of all known ontologies compatible with the data provided by quantum theory is Heisenberg's quantum ontology. The chief features of the world that flow naturally from this ontology are:

1. It is nonlocal; there is some sort of nonseparability of spatially separated parts of the universe.
2. It is both the action and the result of ongoing creation; the fixing of previously unsettled matters is a continuing process. Creativity did not cease with the birth of the universe.
3. It is a natural cohesion of the mental and physical aspects of Nature.
4. It allows meaning: choices can have intentionality, hence meaning.

In every one of these essential aspects the worldview provided by the quantum ontology is the reverse of the one provided by pre-twentieth century science. Consequently, modern science provides man with a vision of himself that is altogether different from the one proclaimed in the name of

Newtonian science. No longer is he reduced to a cog in a giant machine, an impotent witness to a preordained fate in some senseless charade. Rather, he appears, most naturally, within the framework of present-day science, as an aspect of a fundamentally nonseparable universe that is creation itself, both as noun and verb, a creative process that unites in an intelligible way the mental and physical aspects of Nature, and is, moreover, endowed in principle with the capacity to suffuse its evolving form with meaning.

REFERENCES

1. I. Bernard Cohen, *Science and the Founding Fathers* (New York: Norton, 1989).

2. Isaac Newton. Letter to Richard Bentley, 1691.

3. Werner Heisenberg, in *Traditions in Science* (New York: Seabury Press, 1983).

4. Werner Heisenberg, in *Physics and Philosophy*. New York: Harper and Row, 1958, chapter 3. See also David Bohm, *Quantum Theory* (New York: Prentice-Hall, 1951), chapter 8.

5. H.P. Stapp, in *Philosophical Consequences of Quantum Theory*, J. Cushing and E. McMullin, eds. (Notre Dame Ind.: Notre Dame University Press, 1989).

6. William James, *The Principles of Psychology*, Vol. 1. (Reprint: New York: Dover, 1950, p. 147).

SUBJECT INDEX

Action-at-a-distance, 3, 6, 80,
  147, 150, 165-71, 229, 232
Advanced potential, 167
Aether; *see* ether
Air resistance, 46
Alchemy, 17
Algebra, 99-100
Almagest, 2
American Revolution, 4
Analysis, 22, 24, 94-100, 107,
  110; complex, 182
Animal locomotion, 120
Aperiodic behavior, 204
Approximation, 154-69, 181
Approximationism, 168, 174
Archetypes, 129-30
Archimedes' Law, 159
Arcueil, School of, 3
Aspect experiment, 10
Asteroids, 216, 220

Atoms, 3-4
Attraction, 48
Attractor, 183-86, 220; fractal,
  184, 186; strange, 184-88, 210,
  222

Barometer, 105
Basel, School of, 53
Bifurcation, 188-89, 205, 220
Big bang, 231
Biochemistry, 136
Biology, 8-9, 59, 113-22, 131-38,
  180, 182; functional, 137;
  molecular, 133-37;
  evolutionary, 118, 123, 125-138
Biophysics, 136
Blackbody radiation, 4
Block universe, 192
Botany, 100-104
Brain, 242-43; physics of, 243

Nonlocality, 10, 190, 239, 244
Nongravitational forces, 217
Nonlinear: differential equation, 176; dynamics, 175-82, 189, 198-99; systems, 175, 177, 181, 182
Nonlinearity, 7, 10, 175, 207; strong, 239, 244; weak, 189
Number theory, 182
Numerical techniques, 179-80, 190, 198, 215

Observability, 191, 201, 220
One-body problem, 42, 209, 225
Ontology, 228-30, 236; Heisenberg, 239, 241, 244; quantum, 10, 12, 233-34, 237-38, 240-44
Oort cloud, 217
Open system, 205
Optics, 16, 20-21, 102, 107
Orbits: parabolic, 39; elliptical, 47, 160, 163, 215
Organic form, 136
Organization, 136
Oxygen, theory of, 104

Pasigraphy, 104-6
Pendulum, 8, 43, 46, 67-84, 180, 199, 209-10; isochrony, 67, 70-73, 78; nonlinear, 225
Periodic: behavior, 183, 186-87; motion, 184
Perturbation, 53, 157, 189-90, 208, 218-20; stability, 189
Phase diagram, 182-85, 220
Philosophy: experimental, 36, 43; mechanistic, 3, 9, 13, 47, 80, 113, 115, 118-22, 137-38, 176; natural, 22, 24, 54, 60-61, 65, 80, 96-97, 107, 237-38; of science, 154, 177
Phlogiston, 104
Photoelectric effect, 168

Photon, 169, 238
Physical science, 132-34, 138
Physics, 53, 65, 75, 107, 108, 117-20, 131-35, 154-55, 160, 162, 167, 176, 189, 242-43; *see also* dynamics
Pluto, 216, 226
Poetic truth, 172
Poincaré map, 182-83, 184-86
Positivism, 10
Post-Newtonian era, 5
Potentia, 238-42; collapse of, 243
Potential, advanced, 167
Precession, 52
Precision, 65, 74, 91-92
Prediction and predictability, 161-62, 191, 201, 220, 234, 237; statistical, 236, 241
Principia; *see* Newton
Principle of sufficient reason, 242
Projectile motion, *see* motion, projectile
Purpose, 115-22; *see also* teleology

Quantum: jump, 240-42; probability, 238-39
Quantum mechanics; *see* quantum theory
Quantum theory, 2, 4, 9-11, 51, 149, 177-79, 191, 227-30, 233-41, 244; interpretation of, 237; matrix mechanics, 170
Quasiperiodic behavior, 183-86, 220

Radiation theory, 160
Randomness, 176-77, 184-85, 200-205, 220, 236, 241
Rational mechanics; *see* dynamics
Rationality, 153, 156, 160-61, 166-67, 171
Reality, 237

Subject Index

Reductionism, 9, 131, 135; and antireductionism, 134, 136, 190
Refraction, 121
Relativity: special theory of, 11, 148, 171, 173, 227, 232-36; general theory of, 11, 148, 171, 227, 235, 236
Resolution, 94-95
Respiration, 134
Revolution: Newtonian, 21, 28, 30, 44; scientific, 59, 164
Romanticism, 12

Saturn, 52
Scholasticism, 166, 171
Science: history of, 154-56, 159, 174, 177; meaning of, 161; philosophy of; see philosophy of science; sociology of, 154
Self-consistency, 231
Self-reference, 205
Solar system, 41, 52, 114, 116, 148, 214-20; stability, 148; see also; dynamics, celestial
Sound, 210, 225
Space, 235-37; absolute, 228, 233, 235-36; empty, 236
Spacetime, 233, 235, 240-42; absolute, 235
Special theory of relativity; see relativity, special theory of
Specific gravity, 63, 159
Specific heats, 4, 92, 106-7
Spectra, 4
Stability, 217
State vector, 240
Statistical mechanics, 203
Stochastic processes, 205
Strange attractor; see attractor, strange
Strong mixing, 201
Strong nonlinearity, 239, 244
Surface tension, 158
Symbolic dynamics; see dynamics, symbolic

Synthesis, 22, 24, 95-96, 99
System of the World; see Newton

Taylor series, 157
Technology, 227, 231
Teleology, 8-9, 12, 113-34
Teleonomic, 137
Telescope, 121
Temperature, 61
Theology, 229
Three-body systems, 167, 217, 225; see also dynamics
Tides, 39, 48, 49, 211, 213, 214, 225
Time, 32, 35, 65, 240; absolute, 228, 233-36; "Einstein", 240-41; "process", 240-41
Time evolution, chaotic, 203; see also dynamics, nonlinear
Transformation: Galilean, 147; Lorentz, 148
Truth: proximity to, 157; poetic, 172
Turbulence, 180, 188, 210-13; acoustic, 210; incipient, 210
Two-body problem, 42; see also dynamics

Unity-of-plan, 128-30
Universe, 165, 230, 232; block, 192; quantum, 241-42

Verification, 154-55, 157, 161, 163, 165
Vertebrates, 134
Vis inertiae, 50; see also inertia
Vitalism, 176
Vortices; see Cartesian

World-view: mechanical, 114, 117, 138, 176, 228, 231, 236, 239; Newtonian; see Newtonian

252

INDEX OF NAMES

Index of Names

Index of Names

May, Robert M., 222
Mayr, Ernst, 133-38, 143-44
McFarland, J. D., 139
McLaurin, Colin, 167
Mermin, N. David, 223
Mersenne, Marin, 70-71, 79, 85
Michaud, Joseph Francois, 111
Michelson, Albert, 174
Miller, A. I., 169
Monge, Gaspard, 99
Morley, Edward Williams, 174
Morveau, Guyton de, 103

Newton, Isaac, *see* subject index
Nicholson, Marjorie, 176, 221
Niiniluoto, Ilkka, 172

Oldroyd, David, 109-10
Osiander, Andreas, 45
Ospovat, Dov, 138, 141-43
Ott, E., 222
Owen, Richard, 128-30, 143

Pachard, Norman H., 225
Pais, A., 170
Paley, William, 121
Palter, Robert, 57
Pappus, 94, 109
Paul III, 45
Pemberton, Henry, 167
Plato, 109
Poincaré, Henri, 147, 169, 175-76, 179, 182, 199, 206, 208, 220
Poisson, S. D., 3
Pomeau, Y., 188
Popper, Karl, 162-63, 167, 169, 174, 193-94, 197-98, 200, 203, 220, 223-24
Prigogine, Ilya, 205-6, 222, 224
Ptolemy, Claudius, 2, 15

Quetelet, Lambert-Adolphe-Jacques, 44

Ray, John, 139
Redhead, Michael, 170
Remes, Unto, 109
Reyneau, Charles, 90
Riccioli, Giovanni Battista, 72, 79, 85
Ritz, Walter, 165
Robinson, Richard, 109
Roe, Shirley, 138
Roessler, O. E., 222
Rousseau, Jean Jacques, 102-3
Roy, William, 105
Ruse, Michael, 122
Russell, Bertrand, 196, 198-201, 224

Sabra, A. I., 55
Sainte-Beuve, C.-A., 111
Schelling, Frederich, 152
Schlesinger, George N., 197, 224
Schrödinger, Erwin, 170
Schurmann, Reiner, 172
Schwann, Theodor, 120
Schweber, S. S., 140
Settle, Thomas B., 85
Shaw, Robert S., 225
Shea, William, 173
Simpson, George Gaylord, 133-37, 144
Sloan, Phillip R., 111
Smeaton, W. A., 108
Snell, W., 26
Sperry, R. W., 223
St. Hilaire, Etienne Geoffrey, 128, 142
St. Pierre, Bernardin de, 102-3
Stapp, Henry, 7, 10, 12, 245
Steiner, George, 172
Stewart, Dugald, 25
Sturmy, Samuel, 214
Sussman, Gerald Jay, 216, 226
Szabo, Arpad K., 109

Tait, Peter Guthrie, 149, 168, 174
Tannery, Paul, 109

ABOUT THE AUTHORS

I. Bernard Cohen is Emeritus Professor of History of Science at Harvard University, Cambridge, Massachusetts.

Richard S. Westfall is Distinguished Professor Emeritus of History and Philosophy of Science at Indiana University, Bloomington, Indiana.

Thomas L. Hankins is in the Department of History at the University of Washington, Seattle, Washington.

John Beatty is in the Department of Ecology and Behavioral Biology at the University of Minnesota, Minneapolis, Minnesota.

Joseph Agassi is Professor of Philosophy in the Faculty of Humanities at Tel Aviv University, Tel Aviv, Israel, and in the Department of Philosophy at York University, Ontario, Canada.

Henry P. Stapp is a senior scientist at Lawrence Berkeley Laboratory, Berkeley, California.

Frank Durham and *Robert D. Purrington* are in the Department of Physics at Tulane University, New Orleans, Louisiana.